Simon Barnes is the former multi-award-winning chief sportswriter for *The Times*. He is also a nature writer, horseman and the author of more than twenty books, including the bestselling *How to Be a Bad Birdwatcher.* He lives in Norfolk with his family.

Also by Simon Barnes from Simon & Schuster:

Epic: In Search of the Soul of Sport

Rewild Yourself: 23 Spellbinding Ways to Make Nature More Visible

ON THE MARSH

A YEAR SURROUNDED BY WILDNESS AND WET

SIMON BARNES

With contributions by Edmund Barnes

Illustrations by Cindy Lee Wright

SIMON & SCHUSTER

London · New York · Sydney · Toronto · New Delhi

First published in Great Britain by Simon & Schuster UK Ltd, 2019
This edition published in Great Britain by Simon & Schuster UK Ltd, 2021

1 3 5 7 9 10 8 6 4 2

Simon & Schuster UK Ltd
1st Floor
222 Gray's Inn Road
London WC1X 8HB

www.simonandschuster.co.uk
www.simonandschuster.com.au
www.simonandschuster.co.in

Simon & Schuster Australia, Sydney
Simon & Schuster India, New Delhi

A CIP catalogue record for this book is available from the British Library

Paperback ISBN: 978-1-4711-6851-2
eBook ISBN: 978-1-4711-6850-5

Typeset in Palatino by M Rules
Printed in the UK by CPI Group (UK) Ltd, Croydon, CR0 4YY

Where would the world be, once bereft
Of wet and wildness? Let them be left,
O let them be left, wildness and wet;
Long live the weeds and the wilderness yet.

GERARD MANLEY HOPKINS

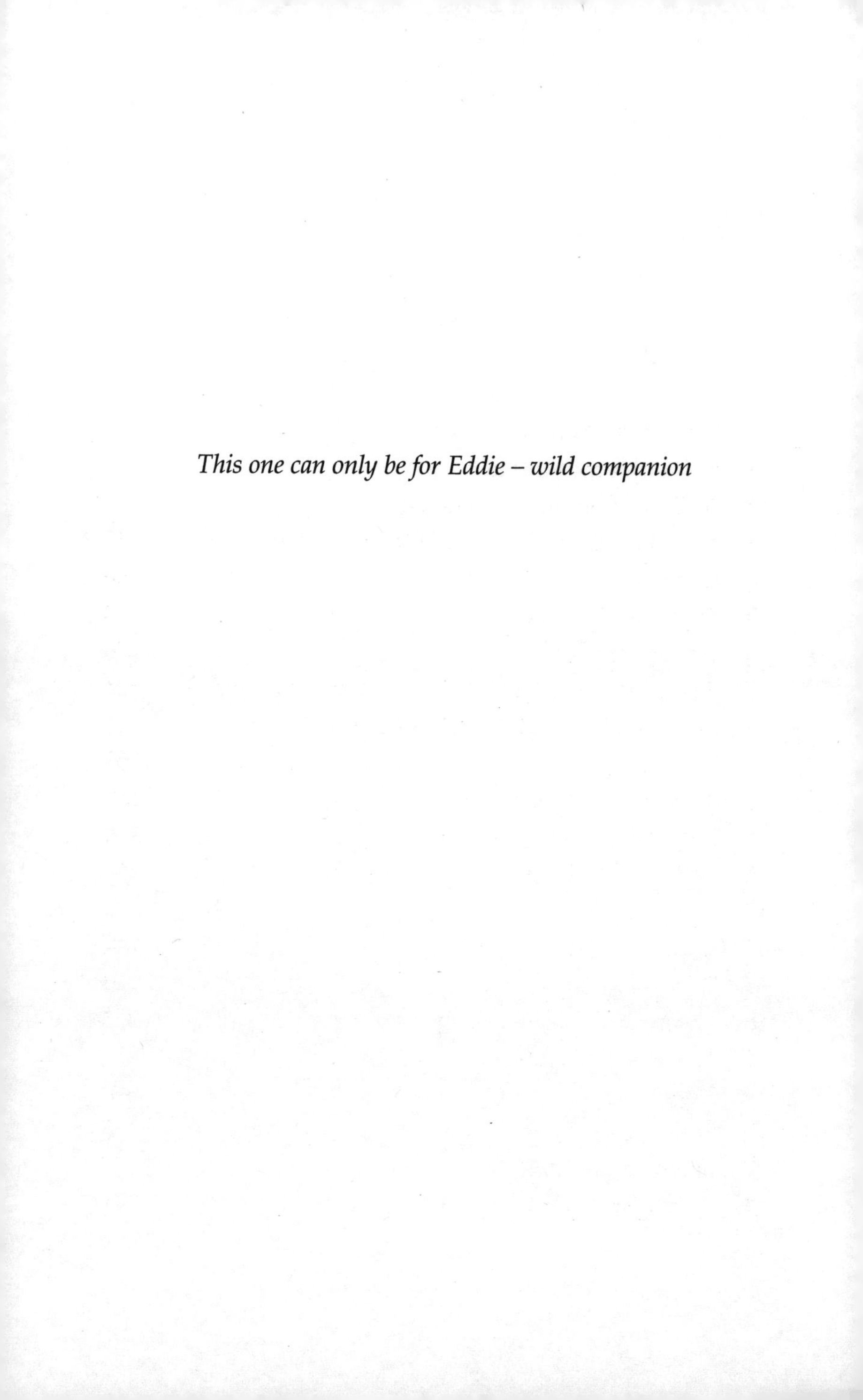

This one can only be for Eddie – wild companion

Contents

	Map of the Marsh	viii
1.	In a Vulnerable Place	1
2.	A Place of Small Importance	21
3.	The Marvellous Nature of Ordinary Things	43
4.	Running Away, Joining Up	60
5.	The Bitterness Test	78
6.	Transformations	100
7.	Being Magnificent . . . Nudge-nudge, Wink-wink	121
8.	Everyone Suddenly Burst out Singing	141
9.	Not Dying	160
10.	Bouncebackability	180
11.	The Ring of Power	200
12.	The Power-ballad of the Earth	218
13.	I am not a Number	237
14.	Runnin' Wild	254
15.	Life is not Tidy	270
16.	The Year Holds its Breath	288
17.	The Mad Conductor	306
18.	Here Hare Here	318
	Epilogue	335
	The Bird List 2012–18	337
	Plant List	341

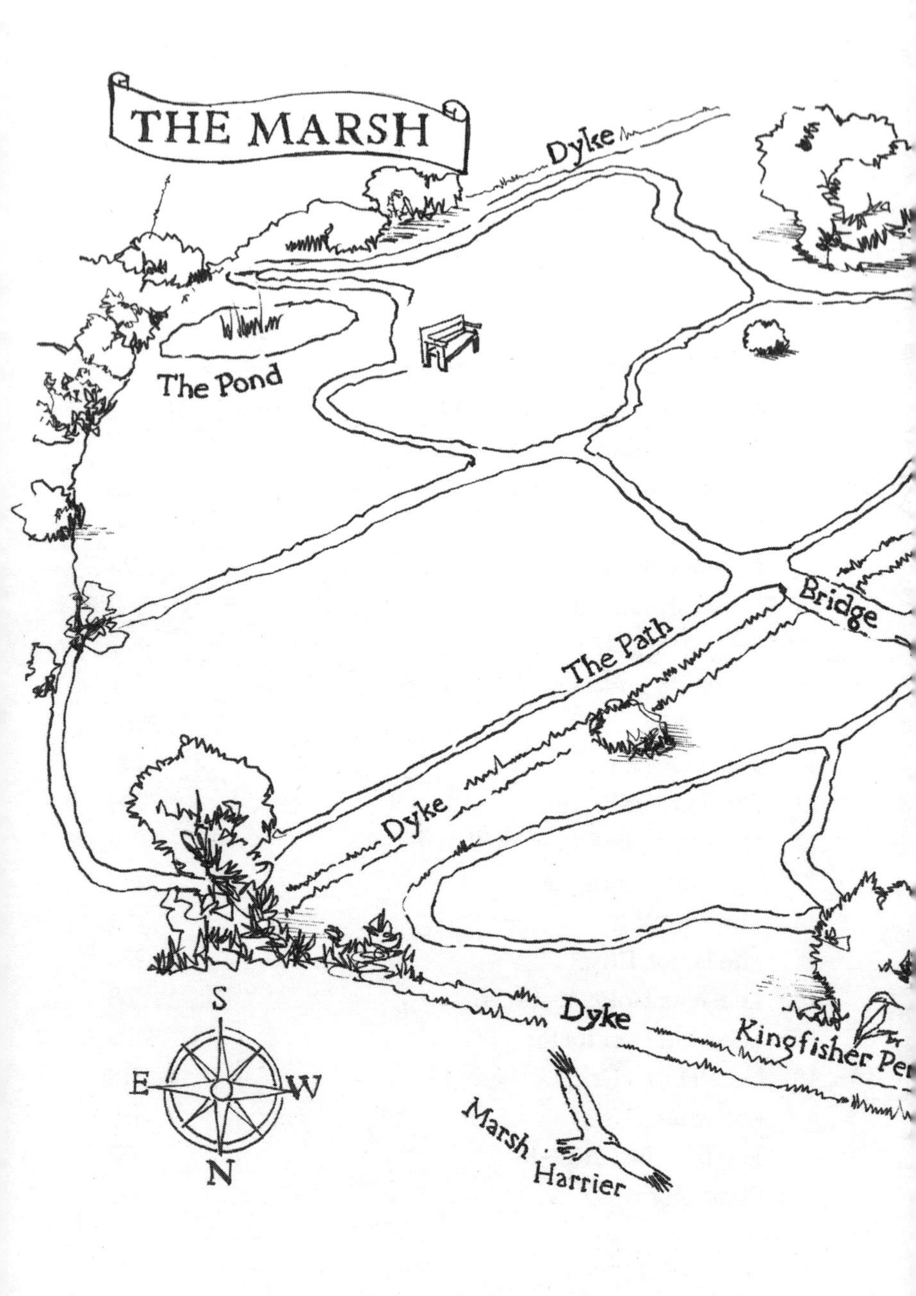
THE MARSH
Dyke
The Pond
Bridge
The Path
Dyke
Dyke
Kingfisher Pe
S
E
W
N
Marsh
Harrier

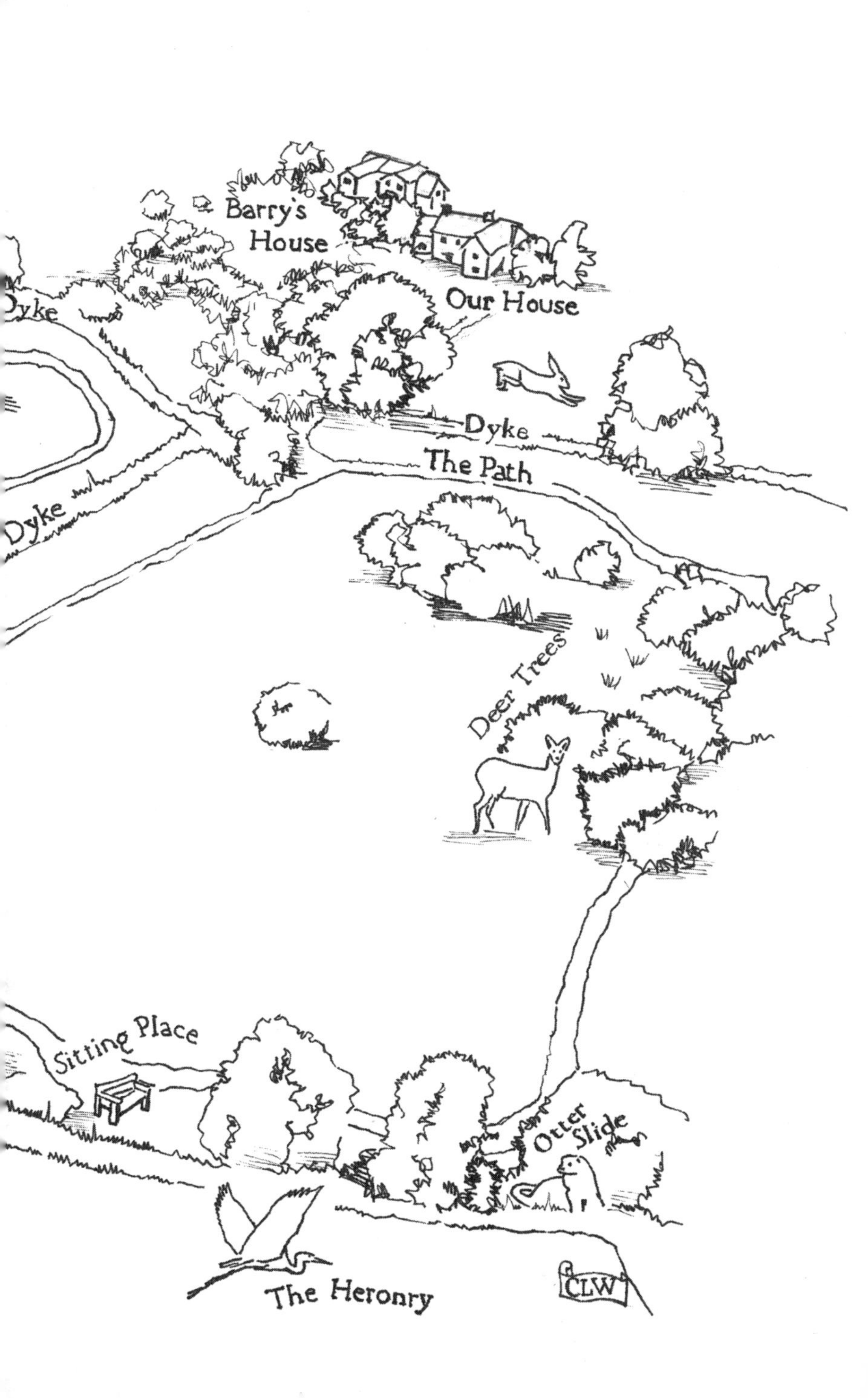
Barry's House
Our House
Dyke
Dyke
The Path
Dyke
Deer Trees
Sitting Place
Otter Slide
The Heronry
CLW

1

In a Vulnerable Place

We were high as a couple of kites – or maybe marsh harriers – so Eddie and I went out to listen to the marsh. There were only us two at home that evening, so it was a bit of a lads' night out. We were high because we'd had a visit from our soon-to-be-ex-neighbour Barry. ''Allo mate!'

He made a great fuss of Eddie, as he always does. Good egg, Barry. Sorry we're losing him but . . .

'Just wanted to let you know that it's gone through, mate. All yours now.'

We'd completed, then. The sale had gone through. But not his house. His marsh. Two or three acres of it. A bit less than two years of negotiation. Relief and joy fought for the mastery. Handshake: Barry is an ex-copper, and with a Barry handshake you always get the idea that something has been settled. Sorted, as he would say himself.

'Thanks, Barry.'

'Cheers, mate.'

The day was already dark, the year was already past the autumn equinox, though only just. I should have been organising Eddie's bedtime, but hell, it's not every day you buy a marsh. So we took a small walk in the dark.

I brought my new toy with us. It's a bat detector: it translates the ultrasonic squeaks of bats into sounds that we humans can hear. When you get good at it you can tell one species from another by the clicks and bonks and splats that come through the hissing speaker. I was no good at all: just smart enough to tell a bat from a cricket. Naming is always good, but it's never everything.

Eddie, my younger son, was 15 then. He loved and loves nature with a sense of driving, inspiring joy. He also loves nature as a thing we share. What does a buzzard eat? Where do ducks sleep? What do bats do in the day? I wonder how many times I read him *Bat Loves the Night* by our friend Nicola Davies. A brilliant book: 'Bat shouts her torch of sound . . .'

'Do you remember how a bat finds his way in the dark?'

'Echolocation.'

A huge word. Articulated with some clarity, too. And to celebrate we picked up a bat: his torch of sound spattering and puttering through the speaker like a bongo drum. Here was a kind of magic: you can't see it, you can't hear it, you don't know it's there, so it doesn't really exist, but turn this little machine on and you don't just find a bat, you seem almost to create a bat. I experienced the wonder of it and I watched the wonder of it in Eddie's face: together we listened to the soundscape of the inaudible world.

We stayed out a little longer, for the night was cool and still. A tawny owl called out rather suddenly, as if he had been working himself up to it for some time. It was a terrible hoot, a truly pathetic attempt at being an owl. This was a young bird setting out to seek his fortune: trying a hoot or two and hoping to discover a hoot-free zone where he would be free to hunt without opposition. You'll have to work on that hoot, mate, if you want to convince the tough old owls that live round here.

Still we lingered. And then a sudden roaring bark, a sound full of anguish and pain, like the hound of the Baskervilles calling for its prey. 'A long, low moan, indescribably sad, swept over the moor. It filled the whole air, and yet it was impossible to say whence it came. From a dull murmur it swelled into a deep roar, and then sank back into a melancholy, throbbing murmur once again ... My blood ran cold.'

It was a truly fearsome sound: one that you'd have thought has been echoing round the marshes of East Anglia for countless centuries, a sound that contained every element of wildness that you'd associate with a wild marsh and the darkness: the fanged spirit that haunts the wet nights of this watery and sinister place.

Eddie knew that one too: 'Chinese water deer!'

It really does have fangs: the males wear long canines rather than horns. And in fact it's a pretty recent arrival.

They've only been around here since the 19th century, released and escaped from private collections and doing very nicely, thanks, on the Broads and in the Fens. Just as well, I suppose: they're declining in their ancestral habitats of East Asia. They're officially classified as Vulnerable.

Vulnerability, eh? There's a lot of it about. That's why we wanted to buy this bit of marsh. Not really to own it, that's a relatively minor pleasure. To stop it being vulnerable.

So many vulnerable things. So many things I wish I could make less vulnerable. Starting with Eddie, who has Down's syndrome.

We were looking at a house on the edge of the Broads when I heard a Cetti's warbler sing out – loud, assertive, unmistakably himself.

'We'll take it.'

An exaggeration, but not by much.

The song of a Cetti's means that the place is wild and wet. The garden's boundary was a dyke, and beyond it there was a marshy no-place that had been left to its own devices. It was runnin' wild, lost control, runnin' wild, mighty bold ... as Marilyn sang in *Some Like It Hot*. The thought of living next to a wild, wet place where a Cetti's warbler sang out, telling the whole world that the place was indeed wild and wet – well, it was at once intoxicating.

The worry, of course, was that the wetland, being vulnerable, would be damaged, destroyed, ploughed, chemicalised, dug up, built on and silenced. So we tried to buy it ... and my wife, Cindy, negotiating with skill and tenacity, succeeded in buying four or five acres of the stuff. We wanted more, of course, you always do, but Barry was able to buy the rest when he moved in. We put up a fence to divide our two chunks of marsh, for that was the deal: one half managed for wildlife, especially for warblers, and the other kept open for Barry's dogs. He also put in a nice little pond. He and his wife loved the place.

But then they decided to move.

> Pigeons on the roof, whispering together. Plotting a coup?

I didn't decide to move away from *The Times*, where I had worked for more than 30 years, writing about sport and wildlife. My departure was involuntary and not altogether easy. Interested readers can learn more with a little internet research, but it's all ancient history now, and there's no profit in raking over it again. What matters in this story is that the Wildlife Trusts turned up trumps. They took trouble over me in troubled times and I'm forever grateful to them – and it's thanks to them that the marsh became a possibility. So in the end, it was wildlife that got the benefit of the trouble taken by the Wildlife Trusts.

The Trusts invited me on a trip to Cornwall, where we went out on a boat looking for dolphins and instead caught a magnificent passage of Manx shearwaters. Then I was asked to go to Alderney, one of the Channel Islands. I was there on the first day of Wimbledon, as it happened. This turned out to be a trip of immense personal significance. Had I still been working for *The Times* I'd have been checking in for a fortnight's work at the Wimbledon tennis tournament. I had scarcely missed a day's play at Wimbledon since the 1980s: great sport and (I know I'll never convince you) rather seriously hard work in sometimes depressing conditions. But instead of taking on a fortnight's hard labour in a long, low, dark room, I found myself scrambling up a rock in the middle of the English Channel, wearing protective glasses and a hard hat, helping to attach satellite tags to half a dozen gannets. I was the wing-clasper. My job was to keep these mighty birds still while the specialist attached the trackers. Another person had the job of keeping that beak – which is like an unusually ferocious spear – wrapped up safe. She used an old sock, pulling it over the eyes so that the bird, finding itself in darkness, ceased to struggle, while she held the bill closed in her other hand. Then all three of us, in a coordinated movement designed to keep the bill away from us, let the gannet go: and each time the bird extended its vast wings, wide as a tall man is tall, and flew.

'Would you like to give this one a name?'

'Yes,' I said. 'Eddie.'

And Eddie Gannet moved away from the rock and circled around it, making the classic cruciform shape. When we had fitted all six transmitters, I sat there for a while as the two conservationists took down all the necessary data. I was within a few feet of nesting gannets, totally surrounded by these bloody great birds, while looking out to sea at many more of the same. I felt as if I was sitting in the vaults of heaven surrounded by angels, not a thought I had often had in the press-room at Wimbledon. It was like being slapped in the face by beauty and wonder and glory, and I was unable to avoid the message that Eddie Gannet and the rest were bringing to me with such glorious and unsubtle insistence.

Enough! Time to move on.

So gannets have become a personal totem of the good

challenges: a bracing call to order, to get on with it, facing forwards towards the future, rather than gazing wistfully in the rear-view mirror. Get on with it. Deal with it. And when in doubt, turn to the wild.

I now have a nice woodcut of gannets on the wall above my desk and a photo of the Alderney gannets as the wallpaper on my phone: twin bracing calls to the present and future joys of wildlife and a prohibition against the temptation to look back at past glories. Self-pity, resentment, anger, hate and nostalgia are all forbidden by the gannets, and they accept no backsliding.

The Trusts then commissioned me to write an ebook about Charles Rothschild. A great man. He was a banker, as Rothschilds tend to be, and a public figure. And he was absolutely nuts about wildlife. He was also a visionary, though what he saw wasn't good. He was one of the first people to see that the wildlife and wild places of Britain were vulnerable. Not only vulnerable but finite. And what's more – what's a good deal more – he was just about the first person in England to do something about it.

In 1912 he founded the Society for the Promotion of Nature Reserves, and four years later, the Society produced a list of 284 sites 'worthy of preservation'. This is now called the Rothschild List. The sites were chosen for two reasons: for their wildlife value and for their vulnerability. It was the beginning of wildlife conservation in Britain.

It was a slow start. In 1916, people had other things on their minds than butterflies, other things than the future of the planet. It was only during the Second World War – and long after Rothschild's suicide in 1923 – that the SPNR and the Rothschild List were accepted as something of genuine importance. In the post-war rebuilding, the SPNR morphed into the Wildlife Trusts, and the Rothschild List showed the way forward. And Rothschild was revealed as the hero, prophet, pioneer and even martyr that he was all along.

It was an honour, then, to celebrate such a man, and to mark the centenary of the Rothschild List with this writing project. I called it *Prophet and Loss*. I began my research with a visit to his favourite place: a little hut in the middle of a marsh. Now it happens that I went on to write the words for the Rothschild project, just as I wrote the words for this book, in a little hut in the middle of a marsh. Different hut, different marsh, same love. Rothschild loved his hut surrounded by wildness and wet, and he went there whenever he could, away from banks and money and public life, and towards the wilder life that we all crave at some level.

Rothschild loved Woodwalton Fen. He bought it because it was vulnerable. He tried to give it to the National Trust to look after, but they turned him down. They said it was 'of interest only to the naturalist'. It's now a National Nature Reserve: owned, that is to say, by you and me, and generously under-funded by the public purse. But it's still a fabulous spot. I had a picnic in Rothschild's hut at Rothschild's table, though I didn't use Rothschild's thunderbox that stood uncompromisingly in the back of the hut. While I was there I saw a marsh harrier from the window – and I bet Rothschild never did. Marsh harriers were shot to extinction by Victorian gamekeepers.

And all of a sudden I had a fantasy. I was right here in the hut with Rothschild 100 years ago, looking out of the window and waiting for him while he got his bug-hunting gear together – and that very moment I had a first for Woodwalton

Fen. 'Look!' (Problem: how should I address him? We could never have been friends, the social and financial distance was far too great, and nor was my expertise enough to bring us closer than social convention would permit. But in dreams and fantasies such matters are infinitely flexible:) 'Charlie! Marsh harrier! Male, flying left to right! Got it?'

'Yes! Far out! Cheers, mate!'

Whatever our differences, the joy – undisguisable, unmistakable, unfakable – of a great wildlife moment would unquestionably have been a shared thing. I have shared a thousand such moments across the world with near-strangers and with close friends; I know such times with immense precision. They're also startlingly intimate, for there is an intimacy with the creature observed as well as an intimacy in the shared joy of sighting it: yes, there, drifting onward, moving beautifully on the very edge of a stall, see how it catches the light and shows off all three colours, silver, russet, black. Marsh harriers came back from extinction during the First World War, but were hammered again by pesticides and reduced to a single pair in Great Britain. Now it's just a part of our landscape, accepted and if still remarkable, no longer headline news. Bloody wonderful, yes? Cheers, mate.

Sometimes I think marsh harriers, recovering from this not-quite-double extinction, by means of their own vigour and by the tenacity of humans who love wildness and wet, are the most significant birds on the British list. They are certainly the most significant birds on our own patch: configuring the seasons and sending us constant messages about the health of the marsh they harry. Without good marshes there are no marsh harriers. Every visit from a marsh harrier tells you that stuff about the marsh that only a real expert on marshes could know. And marsh harriers are experts all right.

I visited five other Rothschild sites as I did the research for *Prophet and Loss.* Two were still perfect, one was a little damaged and overshadowed by industry, one was much diminished but still great and one was almost completely trashed. I wrote about the past and present of conservation: and, of course, the future. The future of the most vulnerable thing of all. You know – the bloody planet.

Whenever Charlie had a little spare cash – which was most of the time – he liked to buy up wonderful chunks of beautiful, valuable and vulnerable wildness. As a result of writing the ebook, I found myself in possession of a little spare cash myself. Well, not exactly spare. A smart person would have used the money to pay off a chunk of mortgage, an increasingly pressing need in these hard times. Or perhaps to upgrade the faithful vehicle that has been carrying us about for a dozen years and more. We needed this money to stabilise ourselves: we really ought to be grown-up and sensible for once. To be totally and brutally frank, we couldn't afford to do this. So many worrying things could be solved with this money. But – well, as you see, we made the other decision. It was an act of knowing folly. Do you think that's wise, sir? The eternal question of Sergeant Wilson to Captain Mainwaring. It all depends on what you mean by wisdom, doesn't it?

Besides, what would Charlie have done?

So we did the same. Cindy did the deal, of course, not me. And Barry accepted.

Handshake.

Thanks, Barry.
Thanks, Charlie.
Cheers, mate.

Morning chores. At the barn I learn that barn owls have the right of way.

At 14.21 on 21 September the earth's Equator passed across the centre of the sun, with the result that all over the planet, day and night – light and dark – were approximately equal in length. The equinox, no less: an appropriate place in which to start, and an appropriate time to acquire two or three acres of Norfolk marsh. Some prefer the term 'equilux': equal light, rather than equal night. Some call it the autumnal equinox, to distinguish it from the vernal variety, but that's pure hemisphere chauvinism. The equinox is a planet-wide event, occurring when the border between night and day is perpendicular. So let's call it the September equinox.

An equinox doesn't feel exactly like a beginning, I admit, unless you're working to the routine of the academic year. It's a milestone: neither a starting nor a finishing point. It's a point where you can stop and eat your sandwiches and admire the distance you have travelled while contemplating the distance you still have to go. And that's appropriate because this business of the marsh was and is all about continuity. We hadn't taken it on in order to change it: we were there to make it carry on. Steady as she goes. We were a safe pair of hands, not a new broom. Our basic job was to keep it safe. To make it Not Vulnerable, which is not the same thing, alas, as invulnerable.

All the same, it's nice to have a sign. A portent. You don't have to believe in such things to rejoice in their significance. It's always pleasing when you find something auspicious: and that's a word derived from 'birdwatching' in Latin. I have always liked the idea of ornithomancy, of understanding crucial things in life from the movements and sounds of birds. It was a serious business: in *The Odyssey* an eagle appears three times carrying a dove in its claws, foretelling the return of Odysseus to Ithaca and his slaying of his wife's suitors. An augur is someone who reads the signs made by birds.

So what would the augurs make of the moment on that day of the September equinox when I went to the barn to collect hay for the horses? I approached the barn – open at one side, with space for a vast double-door that we can't actually afford – and as I was about to cross the threshold a silent white shape passed silently and purposefully over my left shoulder. Not quite brushing against me.

What did this sight mean? Tell me, sad augurs. Was it there to establish an appropriate mood for dealing with the new chunk of marsh? I had often seen barn owls hunting over the marsh – no doubt this individual most of all; and knowing wild creatures as individuals, rather than as members of a species, is a significant thing, as I was to understand with still greater intimacy in the course of the unfolding year. This barn owl had been using the barn as a regular roost for some time, acting all affronted when we paid visits to remove hay, the tractor or Eddie's trike. The darker corners looked a trifle disgusting as a result, because his heaved-up collection of stout black pellets looked horribly like human turds, and there were probably 100 or more of them beneath his favourite perches.

Cindy, who has clever hands, and Eddie had several times

dismantled pellets and teased out the contents: matted fur, pin-thin bones and a sinister little row of skulls: so many deaths in one great vomited slug of unwanted stuff. Posing the question: how many more little lives out there under the rank grasses and the clumps of vegetations? How many shrews on the marsh? How many more shrews than the barn owl needed? Without a vast superfluity, there is no future for a predator. It is a truism of ecology that the abundance of prey controls the numbers of predators, not the other way round. So in a sense, our job on the marsh was not owls but shrews, not shrews but insects, not insects but the plants on which insects feed, not plants but the earth itself. An elemental task then, involving earth, water, fire and air, the fire being the fire of the sun. The barn owls and the other predators were messengers: a top predator is prima facie evidence that everything else in the environment is OK, at least for now.

Dawn and dusk are when barn owls are at their best: not darkness but low light. They love the times of transition. Well, today between hunts he would have the opportunity for an equally long daytime and night-time doze.

This was supposed to be the first day of autumn, but it felt like nothing of the kind. The weather was making specious claims that winter would never come, that late summer would carry on forever: bright sun offering real warmth and a fleet of red admirals parading on the quarter-deck of the ivy flowers. Have you ever noticed that ivy has flowers? It was a relatively new concept for me. No one stops much for an ivy flower: only butterflies.

But to work, to work. I went to my hut and looked out over the marsh. Gannet woodcut above the desk but no thunderbox in the back. Yesterday from my desk I had had a very unsatisfactory glimpse of a large raptor; from my half-second glimpse it looked too dapper to be one of our regulars. What the hell was it? It had flown, dipped, and disappeared behind a particularly annoying blob of a bush. It was no longer Barry's bush: I – or rather we – could make a decision about its future. We could decide all sorts of things.

In Norfolk we call this place a valley and do so without a shred of irony. Half a mile off you can find Hill Farm; a friend visiting from Herefordshire asked why they didn't go the whole hog and call it Lookout Mountain. But out here in the Broads there are traditionally two kinds of pasture: hill and marsh. Our lower pasture is marsh: flat and prone to flooding. The higher one, slopey and easily drained, is therefore hill. Well, by the time you reach the top you are a good ten feet higher than marsh, so where's the joke?

The lower meadow is next door to the garden, which is mostly orchard with a collection of rather odd shrubs; family conferences on their future always end up with them staying, despite my vote for a more bracing outlook. The boundary beyond the garden and the lower meadow is a dyke about six feet wide. Not so very deep, you can see the mud at the bottom when the light is right, but God knows how deep the mud goes. I tried to cross it once: I never found footing and hauled myself out in mild panic with the help of a tree.

The far side of this dyke is where the country goes wild. Wild and wet. This is the marsh proper. It's grazing marsh that's been let go: wet, scrubby, impassable, in many places too soft for easy walking. It's as flat as the lower meadow, for we're on a flood plain which stretches to the river that passes around 600 yards from us, not quite on three sides as it makes a great bend. The river wall has been built up as a sturdy green rampart. Part of me longs to see the river reconnected with the flood plain, a river runnin' wild again. The part of me that owns the house thinks otherwise: we're on the farthest edge of the flood plain and we'd be within the river's range in what the Environment Agency calls an Event. On the far side of the river the plain extends for a while before rising in a gentle slope, which is, of course, the other side of the valley, if those in more lumpy counties can accept the term. To the left there is a stand of trees. The rest is dominated – overwhelmed – by a great stand of sky. We may not do great hills in Norfolk, but when we do sky we do sky.

A Cetti told us to buy as much marsh as we could when we bought the house, but we didn't want to own a Cetti. We wanted a Cetti to own his own place. We wanted to hear that mad shout of song. We wanted to hear that mad shout of song and know that we would hear it again.

So when we moved in we took on the first chunk of marsh covering four or five acres, stretching as far as another fat and treacherous dyke. We set up the fence that divided our bit of marsh from Barry's: a fence that told us that essentially our job was half-done.

Barry's piece is more open than our first chunk, and drier. It has a different feel as a result; the transition wasn't just a matter of fencing. Conservationists talk about 'a mosaic' of habitats; the more subtly different types of habitat, the more biodiverse the place. The richer the place, that means.

But then Barry's piece became part of ours, the fence came down and was re-erected. It now divides our seven or eight acres – savour that, seven or eight actual acres of actual marsh – from our new neighbours' garden. When the Chinese water deer make a mad charge across the marsh they don't have to leap or dodge or – dreadful thought – crash into the wire mesh. They have a clear run at it.

So in the week of the September equinox there was a sense of completion. And with it, a sense of beginning.

> Morning chores. The ivy is a-flower with butterflies.

Paths are full of meaning. The marsh is criss-crossed by paths, but not all of them are easy for humans to follow. The most obvious are those made by the deer. Like most of us, deer have preferred routes from A to B, and the more they walk them, the better the paths become. Use creates use. The paths link feeding spots and resting spots and other important places in cervine life. You can pick them out from what seem like faint shading across the vegetation: slightly darker with use, and following a logic of their own, which is not a logic as humans see it. Often the path has a roof of low trees or curving grasses, more tunnel than road, making it nice and cosy and secure for a deer, but hard going for humans.

Get closer to the ground and you can pick out lower and more subtle paths, used by smaller and still more secretive mammals, some of them carnivores. And more subtle even than these are the paths made by rodents and insectivores, often more like tunnels below the top of the rank grasses and sedges. The short-tailed field voles mark these secret paths with their urine, establishing ownership and reminding them of the way so they can travel along them at improbable speed. The kestrels that visit the marsh can pick out these urine trails because they can see them. They pick up the fluorescence of the urine because they can see ultraviolet light. They are tetrachromatic: they see in four colours, while we humans see only in three, as any television engineer will explain. (Most mammals see in only two colours; primates are the exception. Colour is not an objective fact about the object perceived but about the equipment of the perceiver. A colour-blind human doesn't see it wrong, he just sees it differently.)

Paths are important for humans too. First, they provide access. More than that: paths are welcoming. They invite a human into a place dominated by non-human life. A path makes you feel more comfortable, more at ease in a place. Obviously a path makes it possible to walk from one side to the other without looking at your feet with every footfall. But more than that: it says that humans have a place in this landscape. It passes on a message that the apparent neglect of this landscape is the result of human choice. It's wild, but you have no need to feel alienated. You too, poor human, are welcome here. And in late September there was no sign of the vegetation dying back. Without paths, most of the marsh would be inaccessible or require strenuous effort to get anywhere.

Once a path is established, it gets used. After all, few of us object when life is made a little easier. So the paths that humans make soon become used by all the other local mammals. Drop your eyes every so often: you never know what the next corner will show you.

So when Cindy set off to make a path across the new section of marsh, it was an act of some significance. It was taking ownership, taking responsibility, adding an element of human participation and human pleasure to the life of the marsh. She did so in the toy-tractor we have: a game little Kubota with a cutter on the back. Cindy, as you will gather, is good at stuff. My job was to walk in front, like the man with the red flag, not to slow her down but to fall over any unexpected tree-stumps, tussocks, dips, pits and mounds, rather than waiting for the tractor to tip over.

So round we went, the roar of the engine in my ears and the scent of watermint and meadowsweet in my nostrils: ahead of me tangled banks of vegetations, behind me a new sweet-smelling and inviting path. Cindy, driving with immense competence, even relish, looked like a Soviet poster for the joys of collective farming.

Job done, Cindy put the tractor back in the barn. We then took a bench from the garden out to Barry's pond. Eddie came too, and so did Joseph, our older son, aged 22 and studying music. We brought apple juice for Eddie, San Pellegrino for Joseph – he's an abstemious chap despite paternal example – and for Cindy and me, a bottle of Tesco Premier Cru.

Pop! To Charlie! To the marsh!

In a few moments the sun had done an astonishing thing. It started to set in the east. I'm no expert in these matters, but something told me that was wrong. Perhaps it was the end of the world . . .

It was an illusion, of course, the result of a particularly odd cloudscape. And we're good at cloudscapes here. The clouds in the east had picked up the last few rays of sun, which was now coming across almost at right angles to the flat terrain of the marsh, and they were lit up in a flamboyant array of salmon pink and blue – like a jay's wing – mixed in with improbably lush shades of orange. And all that from the first glass.

A sound from Barry's pond. 'What's that?'

'A moorhen, Eddie.'

'Show me.' So I found a picture of a moorhen on an app on my phone, and played back the call. *Prooop!*

A little owl called, trying to hurry the dusk along. Eddie knew that one all right. Barry's pond was now more of a reedbed. Joseph wondered how easy it would be to remove the colonising reeds. Using all his not inconsiderable strength, he managed to pull out a single clump and then collapsed theatrically. He does quite a lot of things theatrically. It was clear that if we wanted the pond to be a pond, we'd need a digger. And we'd want to take out some of those sallows over there in the winter, wouldn't we? And get some of the scrub cleared. Yes, but not all of it. Maybe if we let the reed develop we'd get breeding reed warbler. Now that's a thought. It was another of the God-decisions you have to make when you are managing any form of land, remembering that doing nothing is also a drastic form of action. You can be managing a window box or a field or a forest or Broads National Park: you have make decisions about what grows and what does not, and what species of animal can live here and what cannot.

Barry's marsh seemed to be giving fresh life to the bit we already had. It was like New Year's Eve: the same old life given a bracing new start. We sat for a while longer, finishing the champagne. One of those rare moments in family life when everything is still.

A furious bark. Eddie correctly identified Chinese water deer.

Sometimes ordinary moments go awfully deep. Eddie wrote a poem when we got back.

on the marsh
it was nearly night
there was white mist
on the grass
it was lovely
we sat on the bench
by the pond
we had crisps

a moorhen house
in the reeds
the moorhen calls near me
and a little owl
a long long way away
maybe in the wood

it was misty cold and wet
I felt excited
the dead tree was dark
against the sky
grey clouds
and black trees
we walked back in the dark
home for a bath

2

A Place of Small Importance

Morning chores and a skylark. Two herons blot out the light.

It's not much, this place. Have to accept that. But then it's also the most important wildlife site in the country. I have to accept that, too.

It's not hard to find a place with greater biodiversity, much greater bioabundance. There are many sites packed to bursting with superstars.

Drive half an hour to Hickling Broad and you can see cranes, gorgeous glamorous six-footers that were extinct as breeding birds in this country for 500 years. Or drive an hour south to Minsmere, where avocets, bitterns and marsh harriers all breed. Head north to Cley and you can find spoonbills and, in winter, sky-darkening flocks of geese. Or go west to Welney, with colossal numbers of whooper swans, or to Strumpshaw, just the far side of the Yare River, where in season there are swallowtail butterflies shining bright yellow and as big as bats. These are wildlife sites in the way that Gerard Manley Hopkins and T. S. Eliot are poets.

Three other sites. All a good bit nearer. I never knew they were there. Beck Meadow, Chedgrave Common and Hillington Low Common. All either in private hands or leased from the parish, and all managed with wildlife in mind, two privately, one by a local wildlife group. Three charming, unassuming places. I visited them with Helen Baczkowska, conservation officer of Norfolk Wildlife Trust. We saw a peregrine flying over one; in another, a grass snake pouring itself away from my boots with what I don't suppose I can call a nice turn of foot. Helen and I dropped to our knees in delight at a turd – well, it was an otter's, so who could fail to rejoice?

None of these places is a superstar. They are, if you like, the holding midfielders, the wicketkeepers, the spear-carriers, the members of the chorus, the corps de ballet – unglamorous but essential parts of something greater. They are called County Wildlife Sites in Norfolk; in other counties they are sometimes called Local Wildlife Sites. They get their designation from the local county Wildlife Trust, and must fulfil criteria that determine their – well, you can't say 'importance' because they're not important. Their relevance, then. They're not really important at all. Merely essential. They work because they are many. They work because they join up the superstar sites and stop them being islands. They soften the brutality of the agricultural countryside. They make room for species other than our own; they make room for life.

They don't have legal status – though legal status is a frail enough protection even for the most extravagantly categorised patches of land. The planning for the mad vanity project of the HS2 train seems to have been a matter of joining up the wildlife sites with the highest level of legal protection – Sites of Special Scientific Interest – much as you join up the dots in a book of puzzles. Why not? After all, there's always less fuss about trashing an ancient woodland than putting a business concern to mild inconvenience. Still, the status of the title of County Wildlife Site does at least – sometimes – add a certain moral force when it comes to questions of planning and can make it easier if you're looking for grants to develop such a site for the benefit of wildlife. There are 1,300 of them in Norfolk: a quiet network that binds together places that people cross the country and the world to visit.

And I was at once filled with a crazed ambition. I wanted our bit of marsh to be a County Wildlife Site. Not to get a grant or to stop anyone building on it, but out of – well, a sense of vanity on behalf of the land itself. I didn't want to say: look, we've got a County Wildlife Site and you haven't. I wanted the land to receive an honour it deserved. I wanted the land to be able to say to *itself*, I'm a County Wildlife Site. I matter.

That's a bit silly, I know. But I felt that the land – well, *deserved* it. So I asked Helen if she would take an informal look at the place. She very kindly accepted.

I have a fascination for people who can do difficult things that I can't do at all. I like to watch mechanics, woodworkers, mathematicians, artists, pilots, musicians, farriers: people wholly confident in an unfamiliar skill. I especially like watching people with wildlife skills beyond my own. It's an education to be with better birders than me. I love being with experts on invertebrates. Also botanists: people who see the land in ways quite different to my own and yet love it in the same way. Birders sometimes refer to botanists as 'stoopers': forever on their knees while we're looking at the heavens. But that's mostly an affectionate tease. It's as if we're looking at different worlds – but it's the same world all along, and it all adds up: that sprig of green I ignore is as important as the distant shape in the sky I am gazing at.

So Helen strolled around the marsh and I watched her botanising with an admiration that's way beyond envying. She found fleabane, marsh chickweed, and was thrilled to find devil's bit scabious, a gorgeous little purple pompom. Also fen bedstraw, meadow vetchling, and as we walked we crushed stems of watermint and its sweet sharp smell filled the air deliciously.

'At first look I'd say it was up there,' Helen said. 'I'll need to get back and do a proper survey in late spring or early summer. But this place looks pretty rich.'

I felt like one of those secret poets with an exercise book full of verse, written alone, pondered over in secret, never telling a soul about their existence and certain that no one else will ever read them ... and then by chance they escape and become known to real poets and real judges of poetry and they say yes, this is *good*. This is the real thing. It may not be Wordsworth or T. S. Eliot, but it's proper poetry.

And that matters. It's the words that count, but being told that you're a real poet makes the words themselves better, richer, more tightly packed with meaning.

This is a nice place. But I wanted it to be a CWS. Packed with meaning.

Chores on a rain-drowned morning. Even the robin's song is a little damp.

Eddie and I took an after-school stroll onto the marsh. Apple juice for him, a beer for me. That sort of stroll. We do it quite often. Get used to it: it will become a recurring theme of this book: beer and apple juice and, often enough, baked beans in a jar. It was still warm enough to do this without being brave: one extra layer each, that was all we needed. On the way to our favourite sitting-place we found a shell collection: coiled and subtly coloured, this one whorled as an emir's turban. We counted them: two dozen in all: houses from which the resident had departed to no good end. The shells themselves were largely undamaged, though a few had a small hole, as if carefully drilled.

What brought them together? The answer should be song thrush, birds who like an escargot or two. They collect the shells in the same place because they use the same stone for bashing them: a classic example of tool-use in non-humans. Tool-use was once considered a trait that separated humans from the rest of creation, but it doesn't. Like all such traits.

There was no stone nearby. No easy explanation: a snail's graveyard; a molluscan long barrow; snailhenge.

Eddie and I took our seats and drank our drinks. A couple of herons flew rather haphazardly over us and around the marsh, lacking the sense of purpose that most herons seem to have with the slow, measured beats of those big arched wings. Youngsters, I bet, still working out what herons are supposed to do. Old term for such birds: hansaw, or sometimes even handsaw. When the wind was southerly, Hamlet could distinguish one from a hawk. So can Eddie and I; we are but mad north-north-west.

One of Eddie's great gifts is for contemplation. Joseph was always ready for the next thing, desperate that life shouldn't escape him even for one second, but Eddie likes to sit still, when the occasion is right. This is not just about lax ligaments and so forth, which are part of the lot of a person with Down's syndrome: he likes stillness for its own sake, when in the right place and the right mood. This evening was one such. I looked at the landscape and scanned the sky for birds; Eddie became part of the landscape. This isn't just sitting like a pudding, though he can do that in other moods and other places. Sometimes out on the marsh he finds a great calmness, which becomes a shared thing, and I'm grateful for it because great calmnesses are not the most obvious part of my life. Eddie shows me as much as I show him.

There were still red admirals on the wing. And a swallow: yes, a lone swallow, southing fast, flying from where we sat at the far end of the marsh towards the house, over it and gone. Every swallow a precious jewel now. The other morning I had seen a bird of prey, but not well enough to get a good ID on it. I thought about that dapper bird of prey I had seen a day or so earlier, obscured by the blob of bush, which was a sallow, the default tree on the marsh. The bird didn't look quite right for a marsh harrier or a buzzard, birds we see all the time. Sometimes a mystery is a good thing, but I would have liked a better look at this one. I kept half an eye open for his return, but in that sweet, slow evening no big bird flew. And it was time for Eddie to get to bed.

'Come,' I said. 'The sky is beginning to bruise and we shall be forced to camp.' He doesn't know the film – *Withnail and I* – or the quote, but he liked the silly voice.

A break in my battle with the inbox. A stray shard of sun has turned a black-headed gull into an angel.

The best thing about running a website is not my irregular blogging or even the hero pictures of me in Africa. It's the occasional emails that come from strangers. Here's a paragraph from one such.

'I took your article about Hickling Broad to heart and my daughter and I visited it in May on our way home from a few recuperating days in Overstrand following my husband's sudden and unexpected death in late January. We only had an hour to spare and alarmed the keen birders with our brisk striding down the paths, but it was the magical place you described. The hides were silent places of joy. Thank you, and more power to your elbow.'

The wild world is good for you. Non-human life is essential to us humans, not just for sustenance but for sanity. Eddie knows that as well as I do, as well as the widow does. But it's easy to be sententious about this. Sure, it's great having seven or eight acres of marshland that I can walk on any time I choose, but that doesn't mean I spend my life in a bird-happy daze. I've known dispiriting days when the soft green acres make little difference, or only in a bad way, when I think of tasks I should have completed, or ways of enjoying it that I've missed out on: a double guilt that can make the marsh feel like one more burden I have to carry.

There's a popular notion that people who have enormous quantities of money could never possibly be unhappy even for a minute. It's presumably wrong, though most of us would be willing to put the theory to the test. So here's one more guilt to throw in: how can life ever be less than ideal when there is so much wonderful nature on my doorstep? I must feel guilty for any passing moment of unhappiness.

But here in this email was a gorgeous and humbling message. The idea that I have helped someone at long range in a bad time is a monstrously life-affirming thing. I wrote back and said something of that. But the fact of the matter is that the way I helped had nothing to do with me, nothing to do with me at all. It wasn't me that mattered: it was what I was pointing at. I had written of Hickling Broad, and of an encounter with cranes. After five centuries of absence – they took part in too many medieval banquets – they dropped in at random to a spot in Norfolk in 1979, birds of incomparable grace and beauty and among my favourite living things on the planet.

But I wasn't just pointing at cranes. I was pointing at all the wonders of the wild world: at all the stuff that lies beyond our own species. It wasn't me that helped to console the widow and her daughter: it was Hickling Broad. It was wildness and wet; it was the places where wild things are. I had the privilege of being the signpost, the pointing finger.

In the same way, this book may not give you poetry that will live with you forever, like T. S. Eliot or Hopkins, but it will point at the wild world – and that, I hope, will be a part of you for the rest of your life, and if I help in the smallest way with that process, then I am deeply honoured to play the part of the pointing finger.

A rushed visit to a lovely spot is not everything you need to cope with bereavement – but all the same, apparently it helped. There are moments of extraordinary elation that come from the wild world: moments of great stimulus and excitement. One of these came when a flight of four cranes flew over the marsh a couple of years back. They paused, performed a slow series of 360s while making the sound of a bugle quartet, dropped a little lower and then, as one bird, they flapped those big wings again and moved on, long necks stuck out in front of them and long, long legs trailing out behind. There are also joys of the contemplative kind, of the kind that Eddie is good at.

But joy is not everything. The wild world doesn't just make life joyous, it also stops life from being worse. Dr Johnson wrote that a great book should help you to enjoy life better or to endure it more steadfastly. The wild world can help you to do both of those things.

In a time of sadness a widow found reasons, not for being less sad, but for living with her sadness. I was a part of that process because I pointed at the wild world. Look to where I'm pointing, then, and find some of life's joy – and also some reasons for living with life's inevitable sadness.

Must get on and write. Not far from my desk a shaggy inkcap awaits.

I'd love to be the sort of person who is so close to the land and what lives on it that I could pluck delicious meals from the fields and hedges any time I chose. But I'm not. At our previous place in Suffolk we regularly had good numbers of field mushrooms – and I was always too busy or too nervous to pick them and cook them and eat them. Which was absurd. My friend Richard, who shod the horses back then, used often to leave our place with his van full of mushrooms. All I had to do was pick the ones next to his, because he knew all right. But I never did. Part of it was simple mycophobia, simple fear of inadvertently eating death's caps instead of mushers. (Well, after all, half a cap is enough to kill an adult.) There's an H. G. Wells story, 'The Purple Pileus', about a henpecked husband who amends his own life by overdosing on mushrooms and returning, maddened, to the family home.

Eddie was helping me to muck out the stables – he has an aged pony called Molly, too old for ridden work, that he's very close to. And we found, between the gate and the muck-heap, what looked at long range like a pair of human skulls. All the more amazing because they weren't there yesterday. They hadn't landed there or been put there: they had grown there.

These were giant puffballs: a wonderfully dramatic bit of landscape. Fungi are neither plants nor animals but are classified in a separate kingdom of their own. That's why people sometimes feel creepy about them: we feel they should act like plants, but they don't because they're nothing of the kind. If anything, they're more closely related to us animals than they are to plants. The stuff that we see above the surface is the fruiting body of a living thing that has most of its being underground in thread-like forms called mycelium. The strange forces that govern their hidden lives had prompted them to create these two vast living lumps right in our field.

We summoned Cindy to admire. We picked one so that Eddie could have his picture taken as he held it: perfect for the picture-diary he and his mother keep – we have volumes going back across the years, read often and then read again. The idea is to help Eddie with his understanding of time and place and his own history, and therefore his own sense of self. In action, they do the same thing for us all, recalling forgotten treats and expeditions that have long gone vague on us, always with pictures of Eddie littler and me younger.

And then Cindy did a fine thing. She took the puffballs in and cooked them. As if we were quite different people: as if we lived here centuries back when there was no longer a Co-op within easy reach and no Sainsbury's and no Organic Man to deliver stuff.

My old friend Ralph grows stuff. His kitchen is an Aladdin's cave of pickles and produce and preserves. Some of it grown, some of it gathered, all of it good. We talk of 'being like Ralph' if we ever attempt anything similar. Ralph also makes the best bread, though I can at least give him a game in that department. It's possible to eat a full meal at Ralph's without a single major item that ever sat on the shelf of a shop. And sometimes I wish I was like that: not just for the food but for the closeness to the earth.

These vast puffballs could have been carved into steaks and fried – Richard Mabey, in his classic *Food for Free*, recommends that you fry them in batter and breadcrumbs – but Cindy went for soup. There was no need for nervousness: you really can't mistake a puffball for anything else. The trick is to pick one that's immature and still white all the way through. Once they start to colour they are, Mabey tells us, indigestible. I thought the soup would probably taste OK, but I was wrong. It was delicious: deep and rich and earthy and creamy: a rare and wonderful delicacy. I'm surprised the shelves of Sainsbury's aren't groaning with the stuff.

'It tastes like heaven,' Eddie said. A remark that made it to Eddie's diary.

> Chores on a long sad morning. Wigeons whistle in the rain.

Every year October comes in with a bang. Damn it.

But let's have a geography lesson. The marsh stands on the flood plain of one of the smaller rivers in the Norfolk Broads. The river passes us rather less than half a mile off, and goes round the two-and-a-bit sides of us in a big right-handed – if you're going downstream – bend. You can make out the tall, grassed river-wall quite easily, standing up from the pasture-land. After they changed their minds about reuniting the river with its flood plain, the Environment Agency decided to reinforce the river walls and create areas where the water can run off and be stored. This can also make you feel nervous: in 2013, the year of the great tidal surge, I remember watching the water cascading down the river wall. And as I did so I found myself looking *up* at two swimming swans. Something not quite right about that. There were fears that the newly strengthened wall might take some damage, and then the river would come hurrying towards us. The next high tide – and it was going to be a seriously good one – was critical. It was due between three and four in the morning.

That made for a nervy night. We created makeshift loose-boxes from straw bales in the barn on the top field – on the hill – because we would need to evacuate the horses if the flood came. There was a lot of getting up and going back to bed again that night, but it all held good.

On the far side of a river there is a flooded area: 100 acres of open water. This holds good numbers of geese and ducks in the winter, also breeding great crested grebes and common terns in summer. A very nice spot. It takes up a good deal of water and holds it, and at the same time provides a home for some great wildlife. What could possibly go wrong?

Here's one of the sad things about life. If you choose to live close to wildlife, you will find yourself in close proximity to people who want to kill as much of it as possible.

Across the river the wildfowlers know neither mercy nor restraint. Such is their right. The sound of gunfire ushers in October. The season of death is upon us.

The difference between an acre and a hectare is that one is a measure of heart and mind and meaning and race memory while the other is unit of area. A hectare is 10,000 square metres; an acre is the area of land that can be ploughed by a yoke of oxen in a day.

Estate agents working in the gracious suburbs are always desperate to tell you that the house stands in a spacious, mature garden of approximately one third of an acre. One third of an acre – buy it and you're not a householder, you're a landowner. You have become a part of landscape, part of history, part of the future. You've no idea how much one third of an acre is, but it sounds about the size of the Serengeti. You have space and green all around you and it's measurable – in acres. You feel as if you could barely walk round the place in a day.

Land comes cheaper out in the countryside. It's an adjustment of scale, not of finance. But you still measure in acres.

An acre measures one furlong by one chain: 22 yards by 220 yards, 66 feet by 660 feet. That can also be expressed as 40 perches by 4 perches; note that a rod, a pole and a perch are all the same length; different terms for the same thing. As I'm sure you've already worked out – in the unlikely event that you didn't know it already – that adds up to 4,840 yards or 1/640 of a square mile. Either way, it's an acre.

That traditional notion of an acre as a long, thin piece of land goes back to the plough: you can plough more quickly and efficiently if you don't have to keep turning. I once drove a horse plough, and while keeping it straight is hard enough in all conscience, it's making your turn that's the real bugger.

But that doesn't help you visualise an acre, does it? You probably don't often use a plough, and only hear about furlongs when you watch the racing. A football pitch has no standard measurement – just minimum and maximum dimensions – but a proper grown-up pitch will usually be around 110 yards by 70. That's about an acre and half. Our patch of marsh is about five football pitches. It's not quite the size of the Serengeti: yet when I watched a barn owl, the one patch of brightness on a murky evening, cross this little bit of land, whispering down onto dinner below, it felt big enough.

All the same, I feel a need to apologise. To talk about acres and ownership can sound as if I'm swanking about immeasurable wealth, especially when there are horses involved. Please rest assured that we can afford horses because we prioritise them; we economise elsewhere. Eddie's pony cost nothing, a second pony cost a few hundred, my own horse a little more. Cheaper than a car anyway, and I haven't got a car. We keep the horses at home, which is cheaper than paying someone to look after them. If we had eight acres of land and three horses in, say, North London we'd have to be shockingly wealthy. As it is, we live in Norfolk, and get by. We lack many of the things that London offers, but we have the wild world on our doorstep, sometimes in the most literal fashion. It's all about choice.

Perhaps I should also add that any impression I might like to give about the perfect marriage and my own brilliant parental gifts should also be set aside at this point. The usual rows – and the other stuff – that mark every marriage won't get much of an airing in these pages, nor will the times when I'm away from home or too busy – or too grumpy – to do stuff with Eddie. The reader will have to imagine them. Not too hard a task, I suspect. Cindy – I should make the point with immense emphasis, hoping it will stay with you, dear reader, until the book's end – has taken on far more of the load of parenting than I have, at the expense of all kinds of personal ambitions and goals. She is the book's most important figure: the sun around which the solar system of marsh and family revolves.

Now, back to the marsh. It's divided and boundaried by dykes, most of them unjumpably wide and fairly deep, especially when it's been raining. The far boundary is maybe a couple of hundred yards – sorry, about a furlong – from the river. The left-hand side of the marsh is wet, very wet in winter, with soggy glades of sallow where the deer like to lurk. It gets drier and more open as you proceed to the right.

The dyke on the furthest boundary runs approximately north-west to south-east. Another big dyke runs straight though the middle, passing through a culvert at the crossing. This is the point at which the tractor-driver needs to keep a good track.

> Morning ride. A weasel crosses our path: what luck! If not for the rabbit.

A pheasant has two ploys for evading predators – for saving its own life, if you like. The first is to lie doggo: to keep very still and quiet and hope to avoid being noticed. But if the predator gets close, the pheasant brings in the second ploy: it leaps into the air with a loud clatter of wings while yelling at the top of its voice. This makes you jump: it's supposed to. It startles you for about half a second, maybe even longer, and that gives the pheasant time to get clear. They are ground-dwelling birds, but they take to the wing to avoid the creatures that can attack them on the ground.

A driven shoot is designed to disable both those defensive ploys. A line of beaters walks across the land, blowing whistles, waving flags that make a great whirring noise and hollering: a deliberately fearsome prospect. The pheasants are unable to lie still in such circumstances: they must take to the air. Once they have run the gamut of their defensive options the guns start to speak.

Every fortnight throughout the season, the local shoot do their stuff on the land adjoining ours. It's very close and very loud: such is their right. So long as they remember to give us notice (there have been a couple of unfortunate and dangerous lapses), we can keep the horses in their stables for the morning, but it's still a pretty exacting experience for them.

Eddie knows what to do on shooting days: he's up and ready and as soon as the guns begin, he's out there to talk to Molly. Molly found trust – at least trust of male humans – a difficult matter until she met Eddie a few years ago. Something in Eddie's uncomplicated affection for her; perhaps also something also in his vulnerability. Horses read human body language better than humans can; it's their primary means of communication, after all. Eddie had spent a lot of time with her, doing work with her on the ground, and riding her till she got too old and stiff for ridden work. His affection for her has continued uninterrupted.

So on a shooting morning, Eddie is at Molly's stable door, a calming presence, saying kind words: 'It's all right, Molly, don't worry. They'll stop soon.' And Molly comes to the door – a pony who once found it hard to give tokens of affection – and nudges him with her big nose.

Fact: they release 40 million pheasants in the countryside every year. Pheasants are introduced birds; in natural circumstances they would be no closer to us than the Black Sea.

There was a sustained hammering of gunfire just behind the stables. I saw a cock bird fly across the horse meadow in a great bustle of wings: a bark from a gun and the bird continued at right angles to his chosen path: straight down. Is that the charm of this strange activity, then? Making a horizontal line into a vertical: at one moment flying as eagles and angels do, the next moment, falling like a pudding, or Lucifer? Perhaps the pleasure of shooting lies in this geometry: in the intersection of one line with another.

Was that the whistle? The whistle that marks the end of the drive? Let's hope so. The cock pheasant was cooling in our meadow.

The figures were in view now, the dogs working the ground with mad enthusiasm. Yes, that was the whistle all right: all over.

Shouldn't there be an element of risk in a blood sport? In any sport worth doing?

But hush.

'Morning!'

'Morning.'

'All right if I—'

'Go ahead.'

'Thanks.'

And the black Lab, fit and busy but still well upholstered as a Labrador should be, entered our field and snuffled about madly for a moment till he found his bird, mouthed it and scampered off towards a great hand full of praise and love.

'You all done now?'

'We'll shoot the other side of the farm after lunch.'

'Have a great rest of day.'

Eddie and I turned the horses out. The season's first shoot was over.

'Well done, Molly. Good girl.' And he undid her head collar, a trick he had recently mastered, and off she went to graze.

'She's a brave girl too, Eddie.'

'I know that.'

The other half of that partnership is not without its brave moments.

There it was again.

Dapper, glidey, more slender than a buzzard.

Stay in view, stay in view!

A half-turn, now cruising at right angles to its former track, caching the sun with a hint of glowing coal, the embers of a fire not quite gone. And the tail, sure enough: deeply forked.

End of mystery.

Assuming it was the same bird, of course, but I bet it was. It was a red kite. The bird was once extinct as a breeding bird in England, pushed back to a diehard population in Wales. Then came a reintroduction programme that began in 1989, and it has been so phenomenally successful that a counterblast has taken place: too many, numbers out of control, reckless project, fostering of glamour species, conservation gone mad, etc etc. Ecologists talk about the carrying capacity of the land: the ability of an environment to sustain a population. More recently, people have begun to talk about the cultural carrying capacity: the number that can live in a place before people turn against them. That backlash has happened, at least to an extent, around the Chilterns, which have become England's red kite heartland.

The kites have spread from there because there's food. That's how ecosystems work. If you provide the right kind of food, the creatures that eat it will thrive. Our bit of marsh provides shrews: barn owls prosper. And kites thrive on death.

There's some artificial feeding going on. Some humans like to see kites and encourage them with butchers' scraps. The Welsh population was artificially fed when it was down on its uppers: I remember years ago watching a dedicated lady hurl a series of grisly morsels to a waiting roost of kites – and then she took a collection from the assembled birders for the upkeep of the local church.

Red kites used to be London birds, back when there was plenty to scavenge in the capital, including the heads of traitors on London Bridge. These days kites do best on roadkill.

And as they spread out from their original reintroduction site, they discover that we don't bother with tarmac in East Anglia. Instead we pave our roads with dead pheasants. Here is an opportunity, and the kites are coming in to take it. They have spread into North Norfolk, where they now breed. There is also a winter roosting site that attracts up to 30 birds at a time. There is pretty convincing evidence that they are breeding in Suffolk. And if they're not already breeding round our part of Norfolk, they will be soon.

Pheasants are reared in pens in vast numbers. They then get turned out into the countryside without a notion of how life is supposed to work, and with no parental example to follow. Pheasants have a great reputation for stupidity: but really they're just naive. They haven't been brought up proper. And they seldom get a chance to grow wiser: if the cars don't get them, the guns will. Never mind: another 40 million will be released into the countryside the following year.

Which is all very strange. Never has a bird profited so much from its ability to die. And now they are helping a long-lost bird to make its return to English life.

There are pheasants nesting on the marsh. I suspect these are birds that have survived more than one season, birds that have acquired experience and wisdom, and so are able to make a good job of raising young without the help of a gamekeeper. All they have to do to keep going is to keep on our side of the fence. There are plenty of pheasants on the far side to tempt that roving kite to stick around.

The kite was not a new bird for the marsh. There had been two or three previous appearances. Which means that our list remained stuck on 99, where it's been for a year or two.

Not that I'm a great lister, for all that lists have their charm for us all, especially birders.

I've recorded – seen or heard – 99 species of bird from our chunk of Norfolk. I count flyovers. There has been some fancy stuff, like the four cranes, and once a European white stork. And plenty of less pretentious species, some of which I've surely overlooked. I once wrote a book called *How To Be a Bad Birdwatcher*, and I continue to adhere to its principles. The Marsh List was into the 90s before I realised I hadn't seen a common gull, which at least made me look at gulls with more energy. I reminded myself that common gulls have wings black-tipped with strong white patches on the black, which are sometimes called mirrors, and added the species to the list within a fortnight. Feeling somewhat embarrassed as I did so.

Some birders only count birds they can see, which makes no sense to me. I'm better when I can't see, to be frank: I don't have a good memory for subtle and difficult visual patterns. I can recognise a reed warbler when it sings and when I see it in the reeds, but I doubt if I could recognise one if it turned up on a fencepost in the middle of a field. Respect (and envy) to those who can.

A bird list represents three things in combination: the skill of the observer, the hours spent observing, and the richness and diversity of the habitat. As I write these words, I find myself making another vain effort to believe that I have heard a water rail from my window, when it was nothing of the kind. A young moorhen, for sure, but one of the things about a list is that it makes you listen and look. You're never off-duty.

And the list (shown in full in the appendix) does show

something of the habitat's nature. A half-decent birder would make a very reasonable picture of the marsh and its surroundings from a quick look at the list. Birds explain place to us, they define it. Like Eddie in contemplative mood, they *are* the landscape.

Perhaps 99 is a bad number, but it's also a thrilling one. It's a number that asks perpetually: what happens next? That's a question you never stop asking if you're tuned in to the wild world. What bird will call next? What bird will fly over? What other treasure will turn up next? And what will happen to the world's wildlife in these troubled times?

There's us and our few acres, anyway. Here we stand, God help us, we can do no other.

All the same – what's going to be the old hundredth?

3

The Marvellous Nature of Ordinary Things

Evening chores. From the Barnes's barn a barn owl erupts.

'Da-a-ad?'

Three syllables, rising to a question mark.

'Ye-e-es?'

'What's a barn owl's power?'

We were sitting on the marsh again, apple juice and beer. Eddie's suggestion: you can't quarrel with his idea of making the most of every last drop of sunshine.

I had seen a barn owl the previous evening. Just like the time before, but slightly different. Well, not much different for the owl but almost shockingly different for me. Every now and then you see a familiar bird as if for the first time: a startling revelation of the marvellousness that is an ineluctable aspect of ordinary things. It is the sort of thing that Basho was able to capture in a few spare syllables. It is the sort of thing Joyce did with his epiphanies, and with everything he wrote. In *Ulysses* an ordinary day becomes a vast heroic epic and there is eternal significance in a cup of Epp's Cocoa, a fragment of seedcake . . .

It was just like that when I went to the brick barn to collect hay for the horses. As I stood in the same doorway the same barn owl shot like a silent bullet over my shoulder. He had been roosting in the barn during the day. Two things made this moment memorable. The first was that it was near dusk and I was looking *up* at the barn owl: the whiteness shone out from the surrounding grey as if the bird was lit from within. The second was silence. Even in this rapid escape-velocity flight, the wings made no noise.

I stood for a moment beneath the flight path, feeling that thrilling touch of closeness: the wild world coming to you, being part of your routine, not as something sought, but as something inescapable and all around you, him getting on with his chores as I was getting on with mine. I disturbed him, sure, but only by a couple of minutes; he was already ready for hunting. They are – savour the fine word – crepuscular creatures, creatures of the margins between night and day. Half-light is their time, the pitchy-black and the stark daylight they swerve.

To share your dusk with a barn owl: well, that was worth a moment's pause. Then back down the hill – is the joke wearing off yet? – to put the hay in the boxes in the right portions for each horse.

As I did so I could hear the soft shriek of barn owl calling to barn owl.

> Morning ride. Blackbird chuckling from the hedges. They don't mean it nastily. Just good nature . . .

In certain moods Eddie loves to ask me an unending series of impossible questions. I've never quite worked out whether he asks them for the sake of asking them or because he knows that if he asks such questions he'll have my attention or because he really wants to know the answer. Bit of all three, I expect.

'Dad – what's a barn owl's power?'

'A barn owl's power, Eddie, is silence.'

I had his attention.

A barn owl can fly without making any sound at all. The wings of most birds make a bit of a din when they're in use; you can hear a swan's wings from more than a half a mile away on a still morning. But a barn owl flies in complete silence. Two good things come from that silence. The first is that the voles can't hear him coming. The second and greater good is that the owl can't hear the sound of his own progress. There's no noise pollution in his ears as he makes his gentle way along the edge of our dykes. That means he can hear the voles, and use the sounds he picks up to locate them with perfect precision. He can hear better than we can, partly because the great disc of his face picks up sounds and amplifies them. I got Eddie to cup his hands behind his ears so that he could hear more clearly the song of a nearby robin.

'And that's what it's like to be a barn owl.'

Sometimes Eddie will fake it when it gets too complicated or I otherwise ask too much of him. At other times the penny drops with an almost audible jingle. He could hear the robin better with his ears cupped: so obviously the barn owl can hear the voles better with his great dished face. I didn't mention the fact that the ears are fitted asymmetrically, one higher than the other. This enables the owl to get a cross-bearing on the vole beneath and so aids the precision. It was complicated enough. Save that for another time.

So we discussed the silent pounce, something Eddie has seen 100 times, the owl's impossibly long white legs extended. 'And what does he have on the end of those legs?'

'Talons!'

'Brilliant!'

This is not the first time I have explained all this. The key to teaching a child with Down's syndrome is repetition. Repetition and encouragement. It's the same with all children, but it's about ten times more important for children with Down's syndrome. You can read that in books. The Down's Syndrome Association will give you information that tells you all that and a good deal more. You can find it out for yourself by trying to teach Eddie anything, from the movement of the stars to the movements of his own body. I wondered why some of his teachers have consistently failed to grasp that principle.

Morning ride. On the distant river I mistake a low flight of swans for a boat.

The Broads is a National Park. But not in the way the Serengeti is a National Park. For a start, it's not owned by the nation. My family, as said before, owns a few acres of it. Lots of other people own much bigger chunks of it. And it's not run entirely for the benefit of wildlife: people grow crops in it, graze cows and sheep, fish its rivers, shoot its skies, live in it, drive cars and boats through it and conduct businesses in it.

The Park, such as it is, extends south across the Waveney into Suffolk, so the title Norfolk Broads is misleading and has been officially discarded. It became a National Park in 1989, since when it's been run by the Broads Authority. This has a remit to look after the wildlife and culture of the place and to help the public to benefit from both. It also has the job of promoting 'social well-being'. It must also labour for the good of the local economy. Not hard to see potential clashes, then.

The Broads is the driest area of Britain in terms of rainfall, but paradoxically it provides Britain's most watery landscape: 125 miles of navigable waterways, seven rivers, including the one not far from my back door, and more than 60 Broads, or areas of open water. Most of these are not natural but formed from medieval peat diggings. The biggest Broads are Hickling, Oulton and Barton. The area includes 28 Sites of Special Scientific Interest, which have cast-iron protection in law from any development or damage – so long as no one important wants to develop or damage them, of course.

The Park extends into a city: the river Wensum flows through Norwich. It includes 2.7 miles of coast. It covers an area of 117 square miles, or 74,872.9 acres, including our few.

There are 6,300 people who live inside the Park, including us; it is visited by eight million people a year, who bring in £568 million. It takes in two counties and six local councils. The highest point is Strumpshaw Hill, which stands a proud 125 feet above sea level; the word hill is woefully inadequate.

There are 15 National Parks in Britain, but the Broads Authority is the only one that has to deal with navigation. The waterways must be kept open if they are still to be navigated, and that involves dredging and reed-cutting. There was a time when the loudest voice among the stakeholders was that of the boating people, who rent craft by the week and by the day. They insisted time and again that the area depends for its livelihood on open waterways, and anything – anything at all that got in the way of this – was economic suicide.

This idea of turning those 125 miles of waterways into something like an aquatic M25 is no longer the central part of Broadlands management. The Broads Authority's publication 'A Strategy and Action Plan for Sustainable Tourism in the Broads 2011–2015' stresses the importance of wildlife. Looking for or at wildlife is the third-most frequent activity among visitors. The report stresses consumers' concern with environmental issues: they want to feel good about visiting the Broads. In other words, the tourism industry needs wildlife whether it likes it or not.

'The Broads is renowned for its biodiversity,' boasts the Broads Authority website. 'It is home to more than a quarter of the rarest wildlife in the UK.' (A quarter of what? Species? Population? Well, never mind.) The visitor may or may not see cranes and bitterns and swallowtail butterflies: but there is still a deep joy in being in the places where such wonderful things have their being. It follows, then, that to get a reputation for the active destruction of wildlife and wild habitats would be disastrous.

How could this frosty morning give birth to a butterfly?

From my desk I watched as a marsh harrier – a female, dark with a creamy head – cruised over the marsh. It disappeared behind the blobby sallow and then reappeared. Damn it, I really must get rid of that bush.

You know, there is no excuse for ever doing a moment's work here. The desk is in the hut overlooking the marsh; I also have a place in the house where I work, but I prefer the hut. Who wouldn't? It's a short walk from the house, a walk that in itself often throws up jolly bits of wildlife, like a late cloud or crowd of red admirals, flying defiantly on into the autumn. Not a bad old commute.

It replaced a more modest hut on the same site, and getting it erected brought us into direct contact with the Broads Authority. Cindy designed it, of course. She sent precise sketches and plans to the planning office. She was at pains to make it clear this was a *nice* hut, one that would look comfortable in the wild and semi-wild landscapes of the Broads. It turned out this was a wrong tack. Kindly but firmly, they said it was altogether too magnificent a notion. They would look more kindly on a something that looked like a cattle-shed. Or a bird-hide.

Cindy came up with a re-draw, a more modest and self-effacing structure, more or less wedge-shaped with a sharply sloping roof heading down towards four windows at the front: a hut that functions as a bird-hide. Is that all right?

Hm, said the planners. It's OK, I suppose. But why are the windows so big? So we can see the bloody birds, she replied, though not in those exact words.

After due thought, they said yes, reminding me of my days on local papers when I frequently had to report that planners had given the green light or the thumbs-up or the go-ahead to some project or other that would make a massive difference to the Borough of Reigate and Banstead.

So as this narrative unwinds and I refer to birds seen from my desk, picture me sitting before a computer screen. Behind it a wall covered in pictures of personal significance: horses, me and my father walking round Cornwall, Africa, that linocut of gannets. On my right, quite a lot of books about wildlife, a picture from the film *Withnail and I* to make me laugh, and a wonderful wooden cut-out of a plunging otter. This last was of course made by Cindy; it carries the words 'Long live the weeds and the wilderness yet' in gloriously weedy calligraphy. Behind me a glass door.

And to my left all the glories of the marsh. That female marsh harrier, she looked as if she was going to stay on here for the winter.

Good. It's what the marsh is for.

'Daa-aaa-aad?'

'Yee-eee-ees?'

'Why are buzzards such good fliers?'

Dad, what's the meaning of life? What is life for? How does it happen? What is it going to do next?

I once wrote a wildlife book for children. It was called – not my title – *Planet Zoo*; I wanted *Unicorn Century*. In either case the subtitle was *100 Animals We Can't Afford to Lose*. In the course of writing one chapter, I realised that before I could get out of it, I would have to explain the problem of inbreeding depression in words a smart 11-year-old could follow. Oh, and in no more than two short sentences or you've lost your audience.

The real problem was not finding the words, though that was hard enough. It was being completely confident that I understood what inbreeding depression actually was. Writing for adults is different: I can assume you understand it, I can write as if you do understand. I can do that even if I don't fully understand it myself. Or if you don't. I can flatter you that you're the sort of person who understands these things, and by doing so, I might even get away with it. I can attempt to bluff because you might be bluffable: plenty of writers work on that principle. We let them get away with it because we're too cool to ask the obvious question. But when you write for children, you have nowhere to hide. Every bluff gets called. If you don't understand it – understand it at some fundamental level – you're buggered.

Eddie's question about buzzards goes straight to the point: how does evolution work? That's easy, son: it works by means of natural selection. That answer might satisfy an adult reader, but it certainly wouldn't satisfy Eddie. So I had to explain how natural selection works, and in terms he could grasp.

The fact that I believe it's possible is, you might say, the founding principle of the Barneses. My father, Edward, was a producer on *Blue Peter* for BBC Television; he went on to invent *John Craven's Newsround*. He passionately and loudly believes that you can explain anything to children. 'Never talk down!' You just avoid grown-up vocabulary – and grown-up pretensions – and tell it straight.

My older sister lectures on the history of art, sometimes to audiences of children, often very young. (I should add that she also lectures to grown-ups on the same subject and in three languages.) She once asked a small Scottish child for his interpretation of the painting of Caravaggio's 'Salome with the Head of Saint John the Baptist' in the National Gallery in London. 'I think he's taking a wee head for his tea.'

It follows, then, that when Eddie asks me a serious question I do what I can to give him a serious answer.

Say a buzzard has four chicks. And one of the chicks is a better flier than the others, so when he grows up he's better at finding food. So this best-flying buzzard finds a female buzzard and they have four chicks. Let's say that one of these chicks is even better at flying than her dad. And she grows up big and strong because she's so good at catching food. When she has chicks of her own, perhaps one of those chicks is still better at flying, and still better at finding food. The buzzards that are the really good fliers will often have chicks that are also really good fliers. The best fliers are the ones that have the best-flying chicks. And so, after years and years and years, all the buzzards we see are really good fliers.

Eddie nodded thoughtfully. For a moment or two he'd got the idea. No doubt he'd ask me the same question again soon enough. No problem there. After all, I wasn't entirely satisfied with my answer. Seemed to me to imply a teleological explanation for evolution, and that would never do.

A great howling mew from above us, and a buzzard rocked back across the sky.

'Good flying,' I said.
'Gliding,' Eddie corrected.
Correctly.

Chores on a desperate morning. The blackbirds are paler than the sky.

What we see is a function of the kind of person we happen to be. That's not mysticism, that's physiology. The eye only receives information: we can't use it until the brain has processed it. A birder inhabits a different visual world to a non-birder. We have trained our brains to respond to birds. Our brains are stocked with visual information about birds.

I remember taking a walk with Keith, a non-birding friend, who was at the time my sports editor. A movement caught my eye: I looked up and flying at quite a height, three birds. Without raising the binoculars I told him: 'Black-headed gulls.'

He made a gesture with his head: one that acknowledged my knowledge and at the same time showed a certain bewilderment as to why I had taken the trouble to acquire it. 'Don't know how you can possibly see the colour of their heads from down here.'

'I can't,' I told him. 'And even if I could, they wouldn't be black.'

It's true. Black-headed gulls don't have black heads in the winter, and even in the summer their heads aren't really black: more like a neat little brown hood. All the same, I could see the birds better than Keith could because my brain was better stocked with visual information about birds.

The subjective nature of seeing fascinates me, perhaps

because I came back to birding after a long time away. Such birding skills as I possess have been painfully acquired – but as a result of this late acquisition, I am aware, at a relatively deep level, that I see the world differently to the way I did in 1981, when I had my Damascene conversion. I hadn't grown up in the same visual world: I made a more or less conscious decision to change it. To change my brain. It's something we can all do, at any stage in our lives, if we wish: I've written books attempting to show how it can be done.

The centre of your vision is the macula. All of us primates have two-eyed front-facing sight that enables us to judge distance with immense precision: that's why lemurs and monkeys and gibbons can jump so accurately from branch to branch. Stereoscopic vision gives us focus, depth and precision. But we also gather information about our environment from way beyond the centre. Peripheral vision stretches out sideways and even a little behind us: say 110 degrees. At these far edges of vision we're not so good at colour, detail and shape – but we're crash-hot at picking up movement. A predator, for example. Or prey.

We're becoming less good at using it. Too much screen time, it's suggested. But here's a thing: if you have anything to do with the wild world, you find yourself constantly picking up movement from the extreme edges of sight.

My writing-hut is perfectly set up for the testing of peripheral vision. My main focus is on the computer screen, but through the window at my left there is a world full of wild movements, and often I find my head moving, turning without conscious decision towards a movement I wasn't aware that I was aware of. Sometimes it's a hunting barn owl, silently demonstrating a barn owl's power.

Today it wasn't.

It moved past in a series of soft bounds: bounce-pouncing its way across the open ground that lies between the hut and the dyke. Glowing orange fur, startling white chest just showing up between the forelegs, fierce, neat little head with teddy-bear ears, and a long tail with a black tip.

Stoat.

I love the stoat stand: up on his hind legs, front paws a-flop, claws facing down, nose higher than his ears, look this way, look that way, in the classic meerkat manner. But stoats are mostly soloists and must look out for themselves: living in that uncomfortable zone in which they can be both predators and prey. Lord, but they can punch above their weight as predators. I once saw a stoat catch hold of a rabbit twice its size and perhaps four times its weight: the rabbit took off, leapt this way, leapt that way, and the stoat was body-slammed from one side to the other a half-dozen times as the struggle spilled over on the tarmac of the road. But the rabbit couldn't break the stoat's hold and hopped, in the end, almost with resignation into cover, where the stoat – still attached – would complete his work.

Here in front of my hut was one of those little vignettes of hidden life. Stoats are not rare, despite legal persecution from gamekeepers, but rarely seen. They don't often go out into the open, they are pretty wary of humans, and they give only scanty clues of their existence. Their long, slender shape makes them adept at hiding in slender places. They are happy to live most of their lives out of sight. You catch one in your peripheral vision and if you have trained yourself to respond to such stimuli, you turn your head and are rewarded, sometimes with an extended glance, as I was that day, and sometimes with an impression that lasts maybe half a second: a fleeting gracile shape, a hint of fast-moving ginger and then a black coda: a coda being literally a tail, as it is also the tail-piece in a piece of music.

Another of those instantaneous dramas played to an audience of one.

Would another person, sitting in that chair, getting on with work with the same degree of concentration, have turned to watch that brief drama? Some would, some wouldn't. It all depends on your brain. And what you've done with it. Would you have turned yourself, dear reader? I hope by the time you finish this book that you will always turn, and so bring more wild things into your life. They're there all right. True, we're losing them at an unacceptable rate, but we're also losing our ability to see what's left. We can do something about both – if we have a mind to.

Music and horses: robin on lead vocal, farrier on percussion.

Phenology is the study of the seasons. It's the science of the way the year moves around us from equinox to solstice to equinox to solstice. The first cuckoo, the first snowdrop: that sort of thing. It's a science of ever-increasing relevance in these troubled and shifting times. And it has a snag: it's easy to respond to the first when it arrives, but harder to respond to the last.

Is that the last butterfly of the year? But every day there was another. Sometimes quite a lot of others, even as the days cooled. Day after day there was yet another last butterfly, and every one of them a red admiral.

Eddie and I have a song that we sing when we see a wood pigeon fly over, often after we've first mistaken it for something more thrilling, like a falcon, even a peregrine . . . anything but a pigeon. But it's a pigeon all right, that epitome of the commonplace, and so we sing a song of our own composition:

It's not exciting!

It's not exciting!

But it probably is, at least in its own mind, as it performs a great territorial glide, flashing its white wing-bars at the world. Our failure to appreciate a pigeon demonstrates only our limitations. Red admirals are certainly commonplace: what if they were impossibly rare? How much would we admire them, that emphatic black, that bloody red, that artful white? And how much we would gloat about the impossibly strong way it flies: in powered straight and level flight they eat up distances with immense confidence, abandoning all the traditional notions of flittering and fluttering butterfly-mindedness. These are insects of power and purpose.

And another day: and another last butterfly, another last red admiral, looking young, fresh and strong, and as if the paint was still wet on its wings. Here is a classic example of the miracles that lie under our noses. Sometimes I think a miracle could take place in our hip-pocket and all we would notice would be an itch in the arse.

To counteract this tendency, I looked up the details of the journey they were about to take. I knew they were migratory butterflies – and that's a hard concept to cope with in itself – but these are very serious travellers indeed. Most of the red admirals we see in spring and early summer are not hatched in this country. They come barrelling in from the continent, belting across the Channel in a northwards flight, coming here to get on with the excellent and necessary task of making more red admirals.

And the red admirals we see at the start of autumn, so bold and so active, have no intention of staying here for the winter; all but the mildest British winters are too much for them. They don't hibernate, as peacock butterflies do.

In their full strength they go hammering south again, back into continental Europe. Often they fly high; radar has caught them rising effortlessly on columns of warm air before exploiting favourable winds to hitch a ride back into central and southern Europe – covering the whole distance in a single journey that takes two or three weeks.

Here, they find that the nettles are recovering after a hot dry summer, and their regrowth is perfect for red admiral caterpillars – and so they breed among the sweet nettles of the southern autumn, allowing the task of making yet more butterflies to continue as it should.

This, then, is a remarkable animal: equipped for making journeys that seem impossibly long and yet capable of all the delicate manoeuvring required for nectar-feeding at the beginning and end of their heroic travels. And yet, because it is common, we seldom stop to celebrate it.

We suffer – all of us, even the best observers, even the most poetic souls – from blindness to the commonplace. The miraculous nature of everyday life is concealed from us. We live surrounded by miracles, but because we see them so often we are sometimes incapable of seeing them at all. The migration of the monarch butterflies the length of North America is a thing of wonder that has been rightly celebrated again and again. The migration of the red admiral is hardly less wonderful a thing: and I, who might claim to know about such stuff, had to look up the details to be sure.

If red admirals turned up in only one or two places in Britain, and for one or two brief weeks, we would worship those strident colours. But as it is, we see them, we note that yet another red admiral has passed by, and then we pass on by ourselves.

Next time you see a red admiral, pause. What a powerhouse this animal is. We use butterflies as images of frailty: who breaks a butterfly upon a wheel? One of the notions of chaos theory – that great events can have tiny causes – is always illustrated with the example of the tornado that was started by the flapping of the wings of a distant butterfly. It's called the Butterfly Effect.

These butterflies didn't look fragile to me as I completed a few autumnal tasks around the stables. They looked as if they could fly through a brick wall. A dozen of them could start a hurricane, easy.

Travelling always fills us with desperate contradictions. Filled with joy, I packed *Roberts' Birds of Southern Africa* and Norman Estes's *Behaviour Guide to African Mammals;* cravenly hoping that the trip would be called off and I would be let off going. Thrilled to be going, I didn't want to leave. I wanted to stay with family, marsh, horses, yet I couldn't wait to step once again into the Luangwa Valley in Zambia, where I was co-leading a trip. And I knew I would arrive back deeply regretting that I had left the beloved Valley, and profoundly grateful to be home, treading once again the paths of the marsh, taking a beer and an apple juice to the favourite bench and singing the Not Exciting song.

But would the butterflies still be there?

4

Running Away, Joining Up

Was it the frost that turned the owl's feathers white?

My body was back on the marsh but my heart and my mind took a later flight back.

Again in my mind I saw that lioness, for she had thoroughly invaded my brain. It was getting dusk. She got up, turning in the course of about five seconds from a hearth rug to the most efficient and thoughtful killer in the Luangwa Valley. Then she went round to each and every member of the pride, which was stretched out across the landscape, occupying an area of about an acre, with each individual lion dozing under a personally chosen bush. She was clearly the alpha female, the pride's chief decision-maker. Sometimes she gave the lion a nudge, sometimes she just walked past, but her gait was elastic and demanding and was obviously, even to us humans, a call to action. One or two – I guessed that they were her own cubs now full grown – got a long lick between the ears. Each lion got up, and stretched – front paws low, elbows on the ground, bum and tail high, just like a cat in front of the cat flap – and began to walk. Well, most of them. One or two of them flopped straight back down again: a lion flop is a fine, dramatic thing; they look like puppets with all their strings cut at once. The alpha female continued her walk, going past the vehicle within stroking distance, and in some moods when telling the tale, I'll swear she rolled her eyes to heaven as she passed me, as if to say: bloody hell, it's like herding bloody *cats*.

Then they were all up. I counted 12 of them, and then ... Well, do you remember the opening sequence of *West Side Story*? The Jets get to their feet, walk across a playground and onto the street. And that's it. But bit by bit, step by step, they move from aimless walk to purposeful dance and the band seems to change step with them, finally blaring out the musical phrase to which the boys will later sing the words: 'The Jets are in gear!' And suddenly they're all dancing, athletic, exuberant, menacing, full of themselves and their joint identity:

Here comes the Jets like a bat out of hell,
Someone gets in our way someone don't feel so well!

And it was like that. Over the course of 16 bars or so, the pride of lions, the Jet Pride – we're the Jets! The greatest! – changed from 12 sleepy individuals into a single unit of power: 12 bodies and 12 brains working as one, the perfect team. It was as if they had just had the greatest team-talk from the greatest-ever manager – and that was the boss, that alpha female. It was that single moment, that moment when the Jet Pride got into gear, that stayed with me above all others on the excellent trip. And the reason it stayed with me was fear. I've spent a lot of time with lions and have had the odd close call: when I see a lion preparing to hunt, it feels personal; I don't feel like a privileged observer, I feel like prey.

Now I was back, looking over another valley, another flood plain, one that had been ironed flat by a quite different river, but one not short of things to love, admire and fear.

Two places united also in their vulnerability. Absurd: a lion vulnerable, an elephant, a herd of buffalo, that hippo that gave us such a fright, all vulnerable. But like the marsh harrier, the barn owl and the deer, they are dependent on the whims of humanity for their survival.

I always get a bit sad and elegiac when I come back from the Luangwa Valley in Zambia. I'll be all right in a minute.

There's a piece of magic that you can do with a horse. Well, it looks like magic, but it's easy enough to pull off. You are on the ground and you put your horse through a series of comparatively simple exercises. When they're done, you turn your back on the horse and walk away. By this time the horse is not attached to anything. Except you. You slow down, speed up, make rapid changes of direction, stop dead – and all the while you have a big soft nose about a foot from your shoulder. The horse is seeing you as leader of the herd of two, and is looking to you for companionship, herd solidarity and moral support.

It's called join-up.

It's about connectivity. So many things in life are about connectivity. Things like managing a few acres for wildlife. It's OK if you make an island of wildness in a sea of concrete, in the manner of Central Park. But it's a great deal better if you can join up.

Coming back from every trip, particularly the good ones, tends to be about counting your blessings rather than wishing you were still there. I was riding Miakoda, my nine-year-old American Paint mare – a blessing if ever there was one – and considering the way things join up. The view from the top deck of a horse, with your head nine feet off the ground, is particularly helpful here.

There's a point on our daily ride where we can look down and across the valley. From here we can see the whole sweep of the landscape, the river snaking its way across the valley floor, and on the far side the wide, shallow waters of the flood: a great wide watery place with stands of trees and herds of grazing cattle.

We then drop down – by this time we have covered more than a mile and a half – to a point a few hundred yards from home. Here again, we can see where it all joins up: all the one flood plain, the ironed-flat grazing marshes that lead to the same river, which lies two or three furlongs away or, if you prefer, 20 or 30 chains.

This tends to be where we roll on into a canter, often startling – and being startled by – Chinese water deer and hares. They hold still until the last moment, just as pheasants do, and break cover right at our feet for maximum startle effect. It's a popular escape plan in the wild world, and the presence of all these mammals – deer, hare, cow, horse, human – again joins up the landscape and unites us in its purpose. It helps – it helps profoundly – that all this land is managed with wildlife in mind by the Raveningham Estate: a practical demonstration that agriculture need not be the enemy of wildlife or human delight in the countryside. Green woodpeckers and jays often shout out as we pass: there's a special and perhaps slightly smug pleasure in this high-speed birding.

And then we loop back up the valley's side, usually at a ground-eating jog-trot, and head towards home, joining up the outward with the inward route.

Back at the yard – untacking, rugging up because it's that time of year already, too cold for naked horses, a thank-you offering of carrot because that's one of those acts of courtesy that keeps the horsey life harmonious, turning out and then getting back to my hut to work – I join up the working and the non-working aspects of the morning.

It's good to know that our few acres are joined up to many surrounding acres in which wildlife is accepted and encouraged. A deer can walk from the estate by way of another stretch of private land onto our patch and then onto the land owned by the parish, the place where our neighbour Jane grazes her sheep: and it can do so in a single unbroken stretch, without any need to cross those dangerous roads, without any need to tell a single human being at all that it's about. You can pick out its paths, accustomed routes across our land and our neighbours' land: land that's joined up by the deer and by everything else. It's easier for a bird, of course: they can get from one place to another by flying over roads and houses, but even for them, it helps to have trees and hedges and rough ground – sources of food and shelter – along the way.

For the wingless invertebrates, the creepers and the slitherers, the connectivity also matters hugely. If you join places up you double their value: one of the reasons why we were so delighted to acquire those extra acres from Barry: we had joined them up – and by joining stuff up you make it less vulnerable. There is robustness in connectivity. Our marsh is not an island. It's part of something bigger. It's part of the stuff that touches and surrounds us. It's part of something that covers the nation: a vast and spreading web of places where the wild things are. And every strand depends, at least to an extent, on all the others: when you break a single strand you weaken the entire web.

We're part of it. Our bit of marsh is part of it. It's part of the web of wild places that unites the local area, the county, the country, the continent, the world. I'd call it the worldwide web, but I think that's been done. And that's a shame because this version of the web is bigger and better and more important than the other, and when we break the strands we do harm to more than just the deer and the hares. We also harm ourselves: for we are connected to non-human life whether we like it or not. With every strand we break, we make every other strand – every other place – slightly more vulnerable. And with it ourselves.

Morning ride. A monstrous bull swan blocks our way. Take me to your Leda!

Only connect.

A good starting point, no matter what you're doing.

If you get into the corner of our tiny outdoor training arena – surface covered in woodchip, surrounded by post-and-rail fencing – you can hoist yourself up and sit in comparative comfort with your feet on the bottom rail. It was a cool, bright day in late autumn and I was doing just that. And watching Eddie lunging his pony.

Eddie was standing in the middle of the school, holding on to one end of a rope while Molly, on the other end, was circling round him at a trot. Eddie was doing all this by himself, with quiet concentration. Connectivity, you see.

Molly trusted Eddie almost at once. Why was that? Did she sense that Eddie was not like everyone else? Horses will often moderate their behaviour for a child. Perhaps she accepted that Eddie was, like herself, a little vulnerable. Either way, her response to him was generous and trusting. First she trusted Eddie, then she got more trusting of me. These days when I went to her box she would keep her head out, and when I raised a hand, she would move her head towards it rather than away. She was a much-patted pony.

Trust is a two-way street. Eddie grew to trust Molly and to feel increasingly confident with her. For some years now, we had been doing exercises on the ground, many of them based on the Parelli techniques of what's called natural horsemanship. So there was Eddie, doing a difficult, complex and – mildly – dangerous job, with calmness and confidence, and he was getting it right.

I taught him how to do it. How very pleasing that is.

I jumped down to help Eddie reverse the direction of Molly's circling, and then pointedly went back to my perch so that Eddie could do the rest himself. Raise the stick, I told him. Don't do anything more, she's already watching you. At the same time, hold the rope away from you and she'll start circling again, in the right direction. And Eddie listened, and Eddie raised the stick the tiniest bit and held the rope away from him and Molly circled, and then he encouraged her into a trot, with his voice and a small gesture of the stick, and she trotted comfortably on, and Eddie stood there, calm and in control.

It was then I remembered a particularly irritating line on one of Eddie's school reports. Mustn't grumble too much – most of Eddie's teachers at the special school he was attending back then were kind and generous and Eddie liked them very much – but, well, one teacher told us: 'Eddie must learn to take instruction.' Or to put that another way, Eddie must learn not to have Down's syndrome. Or to put that yet another way, teacher must learn how to teach. Repetition and encouragement, got that? Well done! Brilliant! Now we'll try it again . . .

Instructions need to be simple. If you give Eddie two instructions at the same time, you'll create confusion: no idea what he's supposed to do, he crashes like a computer, goes into a state of lockdown, won't move forward, won't move back, won't talk. It looks like the most colossal sulk but it isn't: you just have to find a way to reboot. This can be truly annoying, particularly when you know that it's your fault he's gone into lockdown. You've got to switch him off and on again.

But get it right, and you'll find that Eddie's capacity is much greater than you thought.

'OK, Eddie, bring her in.'

And Eddie dropped his stick to the floor, turned away from Molly and went into a semi-crouch. Molly at once went from trot into walk, went up to Eddie and stopped, her nose at his shoulder. Eddie stood and gave her much praise and many pats. A little piece of horsey magic. Join-up. Only connect.

At this point a chunky old 4x4 drove into the yard and from it emerged a tall, big-boned man with large features and the demeanour of a Shakespearean clown. It was Mr Wright from the Internal Drainage Board, and he wanted to know what we wanted to do about the ditches and dykes. So Eddie held Molly while we had a nice chat about access and dates and so forth. The IDB are all about water management: they operate according to water catchments areas rather than local government boundaries. You find them in the Broads, the Fens, the Somerset Levels and one or two other areas. They'll come onto your land and clear the dykes, if you give permission, and charge a very modest fee. The open dykes are good for drainage, but they're also good for all kinds of wildlife including some of our most prized residents – otters, kingfishers, grass snakes. So yes, indeed, go ahead. Water management is also about connectivity.

We humans get plenty of time to study escape ploys. I rather wish we didn't. We are so intrusive in a landscape, so big, usually on the move, and usually noisy. This gives us unrivalled opportunity for the study of different creatures' bums.

It's all right! Relax! Calm down! I come in peace! But off they go, blackbirds flying away with a loud rattle, sparrows scattering in a whirr of wings, rabbits vanishing into burrows and flashing the danger signal of the white bobtail, squirrels skedaddling up trees as if the ground was suddenly red-hot.

Flight-distance is an essential concept. Walking in the bush – as I'd been doing in Zambia – is all about keeping just the right side of a mammal's line of safety. Cross it and the creature's gone. But this line is made of elastic: its degree of stretch depends on the sort of habitat you're in and whether or not the mammal is accustomed to a benign human presence. Get it right and you can walk alarmingly close to many flighty animals. Get it wrong and they're bounding off in all directions, the contradictory mass of leaping bodies designed to confuse your predatory instincts.

You can reduce the flight-distance, at least to an extent, by getting on a horse. You become an honorary quadruped, and you are granted special privileges as a result. I've done that once or twice in Zambia and walked my horse through herds of antelope as if I was just one more of the same kind; I do it most days in England, and ride with Chinese water deer. They have a fondness for the land either side of my favourite path along the flood plain, but they're not herd animals so they're careful about remaining hidden. They can't find safety in numbers so they seek it by lying doggo. Quite often I don't see them at all . . . till they break cover at our feet: the pheasant's startle strategy again, but without the terrible din. I've seen these deer take on a fleet-footed collie at that game: the initial advantage gained by the startle-manoeuvre was never quite overcome. It works. That's the way of things that have been tested by time across the millennia.

Sometimes I've cantered just behind a fleeing deer, and seen it drop its pace, maintaining the same comfortable distance in front of me. Sometimes they'll do this for a mile or so, loping along as we cruise behind. They know there's no need to panic: they have great faith in their own strategy, in the concept of flight-distance.

This time a deer broke cover at our feet. My horse jinked extravagantly and the deer was vanishing at full speed across the sugar beet – no knowing canter this time. As I watched, the deer performed two or three spectacular leaps, its back forming a shallow U. I'd last seen that manoeuvre with an impala in the Luangwa Valley a few days earlier: not a go-faster stratagem but an attempt to show how gloriously fit this particular animal is. Don't bother chasing *me*.

Thus the landscape of home joined up with the landscape of the savannah.

Under a shallow sun a fieldfare quacks welcome to winter.

Quack-ack-ack! A small bunch of them flying overhead, settling in the tops of the low trees, glowing orange-red in the low-angled sun. Here come the fieldfares: crossing and recrossing the marsh, haunting the hedges, restless birds taking to the air and flying off again: thrushes with black tails forever conversing in that triple-quack.

I have never heard a fieldfare sing. They don't sing here, only in their distant nesting lands. For us they are birds of winter: they come flooding in with the cold weather, preferring our balmy climate to the brutal continental winters, arriving from Eastern Europe and Scandinavia.

Perhaps we should hate and fear them. After all, we love the birds that bring in the spring: swallows are birds of good omen and when they appear we rejoice. The breaking of winter's grip, the establishing of high spring: these things are unequivocal goods. Shouldn't we then hate the coming of the winter with the same fervour with which we love the spring? Shouldn't we despise the birds that seem to usher in winter with such blatant quacking enthusiasm?

But we don't. We traditionally fear the creatures of the night: owls and bats have a deeply sinister reputation. Our atavistic fear of darkness goes back to our hunter-gatherer days, when every night brought possibilities of violent death. But winter, as much a killer as any creeping carnivore, doesn't have the same resonance of fear. We humans often fear the dark, but we don't fear the cold. We only hate it.

So as the year turns and the winter thrushes arrive, we don't see them and shudder, knowing that cold days and colder nights will follow their arrival. Rather we welcome them in a spirit of fellow-feeling: like us, they have a winter before them and, like us, their goal is to survive it and reach the sweet spring.

These birds are more comforting than ominous. The year is indeed turning, but we can cope, as we have coped before, as the fieldfares have coped before. We appreciate their gameness in throwing in their lot with us, and wish them good things: mild weather, a superabundance of berries, a safe return journey and a distant nest full of downy chicks come the spring.

In the Luangwa Valley it reached 45 degrees at midday and, of course, there was none of that air-conditioning nonsense.

Some people deal with this better than others. It seems to me that there are two kinds of people: those who fall apart in the heat but are OK at dealing with the cold, and those who are the other way round. Of course, there are always a few who complain bitterly at either end of the scale, and it's a dreadful pity about them.

It always amuses me that, when reporting football, British journalists see cold as a test of manhood and despise effete foreigners who find it hard to cope with football in the middle of a hard frost. He looks great in August, but wait till you see him when it's minus five at Newcastle in February... And yet the same writers forgive British footballers almost any failing when they are forced by cruel circumstances to play in warm weather.

I've always found hot weather easier to deal with than cold. I was relaxed and happy in the Luangwa heat, but when the cold hit East Anglia it seemed proof that the weather and the world were out to get me.

When the weather got serious that year it did so tumultuously and dramatically. You can draw a line down the middle of Britain and make a pretty sound generalisation: on the left-hand side, warmer and wetter; on the right, colder and drier. I sometimes wonder why it is that I have thrown in my lot with the cold side of Britain – and on the bit that sticks out furthest to the east. The wind from that direction comes straight from the Urals without significant interruption, and when it shifts to the north it comes straight from the Pole by special delivery.

I was midway through the transition period from 45 degrees at noon to something nearer five. I was out checking the horses last thing at night – a soothing ritual, at least for me – and looked up to see that absurd dome of stars that we sometimes get here: very little light pollution and, that night, no ghost of a hint of cloud. The stars seemed close enough to reach up and pluck, and there were more white bits in the sky than black. There was already a faint hint of crystalline whiteness about the fallen oak leaves, as I saw in my torch.

'Frost!' I said when I got back inside, rubbing my palms together and rotating my shoulder blades hard.

By morning I realised I had understated the case. This was a virtuoso frost, a bravura demonstration of frostiness, a coloratura frost that set out to show the world just what frostiness could do when it put its mind to it.

I dressed in as many layers as I could while still retaining some kind of movement, and went out to feed the horses. I found that I had exchanged them for a stable of dragons: three twin jets of smoke billowing over the three half-doors, my own hot breath mingling with theirs as I checked that all was well and dealt out bowls of goodness.

The sun was operating as if it had been on a dimmer switch all autumn until now, and the marsh was white enough to hurt the eyes. Not the pure white of snow but a dappled light as each individual colour was picked out and set off by the addition of whiteness. The grass of the meadows was a green made pale by frost; the unwanted pool on the meadow a hard milky disc; the graceful lines of the willow twigs were brown set off with whiteness; further out, the reeds were turning a honey colour beneath the heatless sun and their seed heads were draped in delicate white lace.

I walked out onto the marsh. The frost was so intense it had caused utterly benign plant stems to sprout lines of vicious white spines. The path beneath my feet had gone crunchy. The sun's light was so intense that I seemed to be looking at a landscape of heat rather than cold: the reeds turning orange as I looked towards the sun, the frost on them white fire, smoke pouring from my mouth as I put vigour into my walk to keep warm. Only the vicious thump of the cold – a blow that always seems to land precisely between my shoulders – reminded me of the real temperature. On then to a tangle of vegetation that, for some freakish reasons to do with its open position, had taken the full brunt of the frost and turned stark ghostly white: great tangles of spiked and vindictive pallor that looked as if they might guard the witch's palace in Narnia – I almost expected Maugrim, the chief wolf, to materialise at my feet and ask me why I dared enter.

And then a clump of umbellifers, also surrendering entirely to the frost. There's a kind of firework that hangs in the air before exploding with a dozen streams of fire fizzing out from a central point, and then, at the far end of their brief journey, the lines explode once again. These dead and dried-out stems of cow parsley looked exactly like that: an explosion of frost, frozen in time.

My toes were frozen in my boots. That's enough bloody beauty. Time for a cup of tea.

Thunder from a gin-clear sky. Must be swans ascending.

Never think that cold weather lets you off looking. When it's still colder elsewhere birds can turn up in our icy landscape looking for relief from weather still worse.

It was another shooting Saturday. You have a choice on

these days. You can hunker indoors until the shooting starts, when you pile into your layers of clothing and your boots and come barrelling out of the house full of good intentions but a little late to do the job properly. Or you can stand about outside, stamping your feet, clapping your shoulders and breathing dragon-breaths at the stabled dragons looking over their half-doors.

It's the right thing to do and you know it, but once you've made that decision it always takes the shooters an age to get cracking, while the horses, knowing why they've been kept in, are bracing themselves for trouble.

So Eddie and I were out there in the cold, breathing the sharp, clear air and looking, if a trifle reluctantly, at all the beauty we had been forced to endure as we waited for the mayhem to begin. The delicious super-frost of a few days back had gone, but in this refreeze the ground was hard and the sward before us tinged with white. I looked through the binoculars at three big white birds powering across the space towards us, filling the air with the beat of their wings. Mute swans, of course, though I gave them a hard look, just in case they should choose to be a more glamorous species. How long had my bird list been marooned on 99 species?

'What's that bird, Dad?'

Well, quite often it's a pigeon – not exciting – or a crow, but I turned with goodwill to see what he had found for us. It wasn't as straightforward as I expected: silhouetted against the pale blue of the sky at the summit of a spindly old hawthorn. And the silhouette wasn't wholly familiar: about thrush size but not thrush shape. Still, that doesn't necessarily mean much. Crows and pigeons can often look unfamiliar till they adjust their position and present themselves at a more straightforward angle, especially when their feathers are fluffed up for warmth.

This was no pigeon, this was no crow.

There is a transition that can take place in birding, one that has a glorious drama about it. You go from mystery to wild certainty in the space of half a second. The bird turned sideways and, though still in silhouette, it revealed a bloody great crest.

Waxwing.

One hundred up! Remove your helmet, raise your bat to the pavilion and then to the applauding thousands at the ground and the applauding millions watching on television across the world. A century!

The bird looked restive; I passed the binoculars to Eddie and he managed to get a good look. Or so he said: sometimes he humours me. But I think he got onto this one: my own pleasure in the sighting making this a big occasion for us both. Waxwings come into this country during cold snaps on the continent, often in numbers. They are keen on berries: supermarket car parks planted up with cotoneasters are a regular spot for incongruous irruptions of waxwings; so too are suburban streets with well-berried gardens. They have a thrillingly exotic air about them, a swaggeringly foreign look, a touch of eye make-up and a strong black beard.

The bird hopped and hopped again in the hawthorn, no doubt disappointed at the lack of berries in this ungenerous tree. He took wing: dumpy, rather barrel-like silhouette, pale rump and, just catching the light, a tiny blaze of yellow at the tip of the tail: an extra exoticism, a moment of perfectly timed confirmation. Not much room for doubt here.

And then the whooping of the beaters and the

frighteningly close gunfire all around us, the horses' instant distress, our soothing voices, the pheasants hammering from the lines of maize on the next-door farm where they love to lurk, and, being no fools, heading over our heads and across the meadow to the marsh. They know there's cover there: do they also know there'll be no beaters and no guns? Whatever, as soon as the guns speak, the marsh becomes a bird-magnet and the flight of refugee birds continued until, perhaps 15 minutes later, the blessed relief of the whistle.

When the men and the dogs had finished picking up the fallen birds – none on our land that day – we turned the horses out into the field. Eddie distributed hay in three piles and it was time to go in for a warm.

And a boast.

What does a batsman do after scoring a century? Takes a fresh guard and sets off on the long journey towards a second hundred. No matter how remote, no matter how impossible.

5

The Bitterness Test

Morning chores with overhead geese. Honk honk! Is there a traffic jam in the highway of the sky?

It's sometimes suggested that once you reach a certain age – say 60 – you should be forced to retake your driving test. I have no view on this, not least because I've never taken a driving test. But perhaps at a certain age – say in your mid-60s – you should be forced to take a bitterness test.

Bitterness was on my mind as I took a turn around the marsh on a wintry day that was full of wind and cloud but at least dry. I had received a pair of abusive emails, and they had upset me. Precisely as they were intended to.

They were a response to a book review I had written; the subject was a book about sport. It wasn't a particularly severe review: I've had many worse myself. If anything, the criticism was constructive: it seemed to me that the author had lost the thread of his narrative in his keenness to include every fact and name he had managed to research. I shan't name the sender of these emails because I want to make a general and serious point rather than take revenge.

I had responded to the first message, which was, let us say, intemperate. I had taken a hearty all-writers-together line, and suggested that the only way to shield yourself from criticism is to adopt one of the following strategies:

1. stop writing books
2. stop reading reviews
3. grow a thicker skin

I added chummily that I wished I had managed to keep to these rules myself.

The second message was a curious mixture of fury and gloating. He had a job, while I was no longer chief sportswriter of *The Times*; he, as the proud possessor of a job, was going to drink a glass of champagne (unwritten point: which I surely could no longer afford) and forget my existence. He was going to rise above the futile jibes of a bitter old man. Though I'm not sure that he said 'man'.

Bitter? Was I, though?

I sat on the bench by the dyke, pondering this subject. I had recently written a rave review about a lovely book by a journalist many years younger than me. Emma John's cricketing memoir *Following On* was an original concept splendidly carried through. So I didn't automatically hate all sports books by writers younger than me. What's more, I enjoyed rather than resented good journalism when I came across it and I applauded all good books on wildlife. So OK here in the bitterness department – though perhaps it was an area to keep an eye on.

But if I no longer had the prestige and the financial backing that you get when, say, leading the coverage of a major sporting event for a major newspaper, I didn't have time to get seriously bitter. My freelance practice was absorbing if frequently frustrating, but there's a difference between frustration and bitterness. There can be an overlap, perhaps, so again, keep an eye. But I didn't *feel* bitter. Hell, I had a book to write: you can't get too bitter when writing a book. It might be the best thing you've ever written, after all. Or – less likely – it might be the best book ever written by anybody on any subject. All right, perhaps that's not really possible, but you have to write as if it *was* possible. And that makes prolonged bitterness unsustainable.

You can't stay bitter if you have a horse to ride every morning. You can't be bitter for very long when Eddie is on good form – or when he's on poor form, for that matter. There isn't time, again: you have to deal with what's happening right now. And then, when you're sitting out in this inordinately fabulous place, with incomprehensible quantities of sky facing you—

—what the hell was that?

It was a plop. Something had hit the water about ten yards up the dyke, something hidden from me by a kink in the watercourse and by the height of the reeds. I moved hardly at all. I've got quite good at that over the years: move suddenly and you scare what you're looking for. So I kept my head and my body still and slid my eyes up the dyke. I could just see a couple of ripples, nothing more.

I sat there for a while longer, hoping to solve the mystery. I made a shortlist of suspects:

Water vole. Jumping into the water: they can sometimes be noisy. I had never seen one on the marsh, but had found their droppings – tiny pellets scattered between the stems of the reeds – more than once. And besides, one look at the place tells you they must be there in numbers, the habitat is perfect. Also, there are routinely marsh harriers cruising the dykes and they're not doing it for fun.

Moorhen. These busy and highly splashy birds are always about on the dykes, but they are bold and confident, never furtive. I'd have seen a moorhen, surely, or heard a call.

Mallards. These are also regulars on the dykes, and frequently feed on the surface weeds. But if I'd heard one sound from them, I'd surely have heard a lot more: especially the pattering sound of them filter-feeding, forcing water through their beaks to sieve out the goodness.

Otter. Very possible: a single sound as an otter slithered vigorously into the water. If he was heading away from me, no chance of seeing him or of catching sight of the ripples he created. I have seen otters very close from exactly this bench, and there is a slide kept open by the regular use of the residents.

Kingfisher. Also seen many times on this dyke: occasionally fishing quite unaware of my presence. That's what stillness can do for you. They will make a single, rather hollow plop, emerge – more often with a fish than without – and fly off to eat it at a convenient perch, usually quite a different one from their hunting perch.

So on the whole, that's where my money was. It would have been nice to see it, of course, but it's not essential. I have been breathing the same air as a kingfisher – a possible or perhaps even a probable kingfisher – and that was good enough to being going on with.

Mustn't get self-satisfied, still less smug. Life has its dissatisfactions: that's part of being alive. But bitter? Hell, there are kingfishers in the dyke. Bitterness can be postponed for another day at least.

Cindy is a doer. Idleness bothers her: she always feels she should be doing something, usually something for other people. Her relationship with the marsh is not about contemplation, and seldom about silence. More often than not, when she is out on the marsh she is doing some work there. And right from the start, she was deeply struck by the place's integrity: the way it just was, the way it was its own place, the way it seemed almost hostile to humans who misguidedly believed that we 'owned' it. 'At first I was actually scared,' she said.

That was not just the fear of an alienating atmosphere, a place humans had had nothing to do with for some years. It was also a wholly legitimate physical fear. 'I was scared of driving the tractor into dykes and ditches. Before the paths were established I would be driving into vegetation taller than I was. It was truly wild – unlike the wildness we often talk about – and often almost impenetrable. I had no idea what was ahead and I just had to keep driving forward, hoping for the best. At times I got stuck on fallen trees, but luckily I didn't sink into an unseen dyke.

'The thrilling thing about doing this was that, in all of the noise of the tractor which blotted out the birdsong, there were all the astonishing smells, watermint and meadowsweet in particular. Sometimes when I was cutting the paths in late evening I felt I was invading, that I shouldn't be here with this noisy vehicle. One part of the marsh seemed especially creepy. I remember one evening stopping the cutter and driving as fast as the little tractor would go – back home, leaving the rest of the path uncut.'

The marsh is a powerful place. It's as if the marsh has a strong sense of self, like certain powerful personalities who – garrulous or silent – dominate any gathering just by being themselves.

> Another piece written. Modestly I acknowledge the applause of a flight of swans.

There is a mighty ash tree on the bank of the dyke that divides the garden from the marsh. I can see it from my bed: in the leafless seasons I can make out in its branches the features of D. H. Lawrence and Charles Darwin. It shows no sign of Chalara, the dreaded ash die-back disease; long may that stay true. My writing hut stands 15 yards from its impressive trunk, around which two people could just about touch hands. The trunk rises for ten feet before dividing into three, each separate course making its own spirited attempt to reach the sky. It follows that the tree has been in the way of many glassy stares between sentences as I write in my hut.

It's movement that attracts the eye. You are more likely to escape notice if you keep still: and throughout nature this valuable piece of information is shared by both hunter and hunted, as we've already seen in these pages. And from the tail of my eye – peripheral vision again – I was aware of movement on the trunk. I turned and looked and there was as charming a piece of magic as you can see in Britain or anywhere else for that matter.

On the trunk, two treecreepers. I had seen treecreepers at our place several times, always and only on that tree. Once I was with Eddie, and he was delighted by this odd and special bird: a bird doing its damnedest to become a mouse as it crawled in purposeful darts across the acreage of bark. In the first weeks after we had moved in, I heard treecreepers out of sight, lost high in the deep foliage of the ash. Now I had them in sight again: round and round they went, each performing a spiral ascent of the trunk, using those thin, curved beaks to find tiny fragments of life in the bark. They are not vastly rare, but they're hard to see, being mostly birds of the canopy. That sweet, thin, high voice is normally the only clue: sometimes a jumbled collection of notes as a song, more often as a single sustained note that's used for contact and alarm.

I watched them for a good five minutes. Then they vanished, heading back up to the canopy, perhaps to work their way down again. And I remembered that I hadn't marked the waxwing on our Official Bird List. So I got it out from my desk drawer and turned through all the names of all those British birds. Many of them I would never see on the marsh: Egyptian vulture, Hudsonian godwit, ancient murrelet, dark-eyed junco, magnolia warbler. I wondered how many species on the list had visited the marsh without my being aware of them: birds a crash-hot birder like my friend Carl would have rejoiced in.

As I was scanning the list for waxwing, my eye passed over treecreeper. And I found, to my surprise, that it had not received its highlighting with the green highlighter. I performed this task, and then did the same for waxwing. Then I counted, and recounted. Yes: 101. I had been on 100 birds for the past year or more without noticing. So there's a parable: what you seek is often already achieved, already all around you, if you only had the wit to realise it.

It was also confirmation, if it were needed, that I really am truly appalling when it comes to figures. But that meant we were overdue a celebration. A bottle of Co-op champagne, perhaps. I thought it would be inelegant to inform my email correspondent – he of the muddied narrative – of the fact.

The champagne was not bitter.

Jake regards himself, not without pride, as the black sheep of his family. The family in question is called Fiennes; his brothers and sisters are all involved in acting and films and music and being famous. Jake is the creative one. Certainly he is as brilliant and as creative as any of his siblings, but his medium is different. Jake works in land.

He is manager of the Raveningham Estate that runs not quite next door to our place. The place is owned by Sir

Nicholas Bacon, a member of the landed gentry. But not in the diehard old-fashioned sense: he employs Jake, and Jake runs the land for wildlife as well as for profit. The crops are surrounded by high and hairy field margins, full of wildflowers in season – and therefore full of pollinating insects. These generous spaces are surrounded by lofty hedges. Their trimming is timed to produce the maximum crop of berries and other food for overwintering birds and at the same time to provide maximum protection for the springtime nesting. Woodland areas are in season deafening with song.

Jake and I were bouncing across the marshes of the estate. Jake's Land Rover was doing the hard work; I was just holding on. Jake was talking about his time as a jackaroo on an Australian cattle station, and how he had to shoe his own horses. Get it wrong and the horse goes lame and you don't work.

We parked on a high ridge – a bund – in the middle of the marsh. We were the loftiest things for many miles around. The light was fading fast, but I could still pick out the lapwings that had come to rest for the night: white faces and white flanks allowing them to stand out in the increasing dark. Occasionally one would take off and make a brisk, floppy-winged flight over the others to look for a slightly better, slightly safer place. The occasional reedy call.

I may not be the best at accountancy, as the treecreepers will tell you, but I like to count birds. Especially, I like to count birds in numbers. Count to ten: then count in blocks of ten. And in this case, when you reach a hundred, count in blocks of a hundred. That made around a thousand lapwings.

Lapwings do things to my heart. They lift it and they break

it more or less at the same time. I remember seeing lapwings in thousands as a boy and thinking nothing of it: oh yes, peewits, jolly good, anything more interesting out there? Now, like most birds that have anything to do with farmland, they have declined horrifically. But here they were present in decent numbers, and that made me happy on a relatively deep level. It seemed to me that there was hope and despair side by side in front of us: and as the lapwing called, it even seemed permissible to hope for victory.

We sat on into the dark. Jake is not a birder in any conventional sense of the term. He is more of a lander, I suppose. Good numbers of good birds on his land are proof that he's getting the land right.

I seem to have spent quite a lot of time sitting in the darkness with Jake in his unheated Land Rover with the windows open, and generally a serious bit of wind blowing in from the Urals. We talked, we were silent, we talked again. And then—

'I can hear them.'

A moment later so could I: a honking that was gentle and conversational rather than strident and klaxonish. And then, at once, the blank page of the sky was inscribed with the most perfect calligraphy: great abstract swashes and arabesques, lines that zigged and zagged in a series of shallow vees, and with variations on a theme of vee.

Pink-footed geese. Pinkfeet.

They come in to join us from Greenland and Iceland, and they choose the bleakest places on our own island to overwinter. Once here, their daily routine is to fly from their feeding places to their communal roost; they're usually the last of the daytime birds to go to bed. There they were, in their chattering straggles, filling the sky with noise and their own staggering visual presence.

That's ten. That's a hundred, a hundred, a hundred ... Once again I counted a thousand as they settled into the marsh and disappeared into the darkness, still audibly discussing the salient points of their journey and the merits of their overnighting spot.

By this time we could barely see the bonnet of the Land Rover and my fingers had been welded onto the binoculars by the wind.

'Do you think it's time for a beer?'

As a matter of fact, I did.

Neck cracking, head turning at wagtail alarm call. Sparrowhawk! Two half-decent birdwatchers, then.

We still called then cigarette cards – more often fag-cards – but by the late 1950s they were more likely to have come from packets of tea. With a quarter-pound of Brooke Bond tea – loose tea, none of that teabag nonsense – you got a fag-card. It would have diminished their power to call them tea-cards. Mostly they were coloured pictures of wild creatures. They were things to be treasured.

The playground of Sunnyhill School would every so often become a gambling den, the Las Vegas of SW16. During these times when the craze was running hot and strong, every boy – girls did not take part – was mad for fag-cards. And while there was a fair amount of swapping and general trading, the tone was set by gambling. You set a card up against the wall as a target: any player might flick a card at it, and if he knocked the target card over, he kept it. If he missed, the stallholder kept the card in question. The odds were substantially in favour of the stallholder: here was a Marxist sermon, or perhaps one from the other side. The stallholder had even more of an advantage with Bombsies, in which skill was largely factored out of the game. You dropped a fag-card from a height and hoped it would land on top of a card laid on the ground. Sometimes the dropped cards were pierced with a hole punch in an effort to increase accuracy; these were often refused by the stallholder, as being a crooked practice. (There were occasional attempts to rig the flicking game by sticking together two identical cards, thus creating a more formidable missile.)

I seldom if ever took part in these games, for they went against my temperament. But I loved the cards themselves. I succeeded – rare thing – in persuading my mother to alter her habits: she agreed to purchase Brooke Bond tea. I never had the full set of British Wildlife – though I eventually gathered all 50 Tropical Birds – but I got most of them.

And what a revelation they were. All those mammals I had never seen. I was able to accept the idea of pine marten and wildcat because they came from Scotland and that might just as well have been the plains of Africa or the surface of the moon. But the creature that truly baffled me was the stoat.

I knew about stoats because I had read *The Wind in the Willows* many times over, and I was familiar with the illustrations of heavily armed stoats by the great E.H. Shepard. But the suggestion, embodied in these fag-cards, that stoats were creatures that ordinary people might actually set eyes on . . . well, that was absurd. Preposterous.

Few children think things out to their logical conclusion.

It seemed to me then that stoats were at the same time fictional and non-fictional creatures. They had an existence in the eyes of certain specially qualified people, but not for me. They were beyond my scope. For me at least, they weren't quite real. They were like unicorns.

I granted stoats a sort of shadow-reality. I accepted that they weren't purely imaginary, like dragons, but I knew they weren't creatures you actually saw – not like the pigeons and blackbirds and the occasional cuckoo that, back in those days, could still be found in the woods at the top of Streatham Common.

I suspect many of us accept the same kind of limbo and bring it into adulthood. There are creatures that aren't obviously mythical, in the manner of griffins and hippogriffs and harpies, but which are nonetheless beyond our scope. It was much later in life that I discovered that you can summon many of these creatures from limbo and revel in their presence – and it was the great discovery of my life. Just about every line I have ever written about wildlife is written in the light of that discovery: so that you, dear reader, might also learn how to bring these semi-mythical creatures out of limbo and have them leap and dance and sing before your own living eyes.

For I have seen stoats many times. Never without surprise and delight, always with that slight sense of seeing a mythical beast come to life. And then, on the marsh, came the most miraculous stoat sighting. I had her in view – not sure that it was a female, but I'm not happy about using 'it' for so confiding a creature – for several minutes. I then made a note: *whoreson mad stoat*. The gravedigger in Hamlet referred to Yorrick as 'a whoreson mad fellow': well, this stoat looked madder.

She danced for me, this improbable creature. Stoats and weasels are long and slinky and snaky: if they have a black tip to their tails they are stoats; if not, weasels. Both are fierce predators that bite well above their weight: they can not only kill a rabbit but they can pull the corpse into cover as well, which is a bit like a human doing the same thing to a cow.

There is something compelling about a stoat: the bright intelligence that shines from the eyes tells us at once that we are watching a fellow mammal at work. Fellow feeling is part of any meaningful encounter with a stoat. They cross the ground – when not hurrying – in a manner that reminds me of one of the exercises I did when I was trying to learn italic writing: *nnnnn*.

And then the stoat stood up on her hind legs, making herself about six times taller than she was on all fours, stoats being low to the ground. Once she had attained this dizzy altitude, she did something utterly bewildering: she threw herself on her back, writhed like a man drying his back with a towel, and then jumped onto all fours again, whereupon she leapt into the air, twisted, and again landed on her back. The whole thing was shockingly unexpected. The dance continued for maybe three minutes: she looked like a furry snake, like the great python in *The Jungle Book* that fascinates the monkey-people with The Dance of the Hunger of Kaa.

Then she stopped abruptly and looked all round, no doubt scanning for potential predators, but with the air of an over-stressed passenger at an airport who would welcome a row at the least opportunity.

And then she was gone: as if she had just willed herself into the landscape, or the landscape of the marsh had chosen to wake up and swallow this tempting little morsel.

What the hell was going on? There seem to be two theories. One is that the stoat was dancing to 'fascinate' a rabbit: to dazzle potential prey and reduce it to a state when it doesn't know how to react. The fact that there was no rabbit doesn't entirely negate that theory: it may have been practising, just as a pet dog will pretend to hunt a rabbit.

There is another theory, one with rather less charm, which is that the stoat is suffering from parasite called *Skrjabingylus* that lodges in the brain. It was a grim thought: but parasites too are part of the vast, tottering, improbable structure of life. Without them our world would have turned out very differently: it's reckoned that parasites are one of the most potent drivers of evolution. We can speculate that stoats would not be stoats and humans would not be humans without parasites. We should celebrate these creatures that live in guts and brains and blood and eyeballs: and do so because they know no other way to live.

Twelve blackbirds foraging in the field.
Half a pie.

As autumn moves towards winter the blackbirds start appearing in the garden and in the bottom meadow. The garden holds about a dozen apple trees, not a fruit I have much enthusiasm for. Cindy, as is her nature, has a boom-and-bust relationship with them: sometimes gathering them up and giving Eddie powerful breakfasts of porridge with hot stewed apples; at other times, when there are more pressing concerns, leaving them on the ground. That's what the blackbirds prefer.

At any time during this season of fallen fruits, there will

be half a dozen blackbirds feeding on them energetically. And here's an odd thing: every one of them a cock. There they are, sleek, glossy and banana-billed, pecking fiercely at the fruit in a great overflowing of male solidarity. You see an occasional female, dark matt brown and with a dark beak, holding her own and getting her fair share of fruit – there's plenty to go round – but mostly these are all boys together, as if they were at a particularly seedy pub. Perhaps they even get a little drunk on the fermenting fruit.

When they're not under the apple trees they'll be ground-feeding on the meadow, operating in that on-again, off-again style of blackbirds: a long pause to listen, perfect stillness, then a few rapid hops to a new position – and every now and then a moment of triumph, digging that yellow bill into the soft earth in a wriggling moment of achievement. I saw them most mornings when I was feeding the horses, and usually made a count. And it was 20. It was almost always 20: a gentlemen's club with a fixed membership – got to keep it exclusive. That tends to be the way with any limited resource. The far corners of the meadow, and in the middle along the fenceline: these seem to be the favoured areas.

I usually remembered to bring the binoculars for this morning ritual, and usually cast an eye over the blackbirds, partly for the counting, and also to see if one of them might turn out to be a ring ouzel. These are blackbirds with clerical collars: same genus and all that, but birds associated with high country. They turn up every now and then in East Anglia all the same: but never here, not in our time. But here's a thing: the checking itself is a part of the pleasure – pleasures of awareness, of involvement with this patch of land. The tiny disappointment at the shortage of ring ouzels is no disappointment at all: it's a way of letting the ring ouzels know that if they ever turn up, I'll be ready.

Back in the house, cup of tea, morning paper, wincing at the impacted clichés of the sports pages (bitter? Me?) and noting that the blackbirds are hard at it beneath the apple trees. They're not all cocks by mere chance. This is part of a fiercely calculated survival strategy. Because it's not just about staying alive: what's the point in staying alive a little longer if you leave no descendants? Most of these blackbirds are not British: like the fieldfares they have travelled from Scandinavia and continental Europe to take advantage of our mild winters. You can't go digging for worms when there's a hard frost: and when the frost is winter-long, to tarry is to starve. So these blackbirds take the time-honoured avian tactic of coping with crisis by being somewhere else when it happens. It's called migration.

Most of the blackbirds that breed in Britain stay put for the winter, but that doesn't mean all the blackbirds in the world follow the same plan. You mustn't think of a species as a single entity: identical birds all doing identical things. In different places, different birds of the same species act in quite different ways. When I lived in Asia, on an island outside Hong Kong, the sparrows that played and chirruped around my house – I had the ground floor and the 'garden' – were not house sparrows, as they would have been in Britain, but tree sparrows, which in Britain are entirely country birds.

Most of the blackbirds at our place in the autumn were migrants: transients, as opposed to residents. But why all male? What happens to the females? They turn their beaks up at the things our climate has to offer, and carry on, wintering in southern Europe where the food resources and the climate are more reliable. So why don't the cocks go there as well?

They want to be nearer home. As soon as the breeding grounds have unfrozen themselves, the cock birds want to be back there staking their claim. They want to be set up and singing, fully established in a fine territory of their own, by the time the females get back from their extended winter break. And while it's true that in theory a really effective and experienced male should be able to turn up at any time and help himself to whatever territory he chooses, the fact is that a male defending a home territory has all the advantages over an interloper. Home advantage is a potent force in the struggle to become an ancestor – a truth that is re-enacted every Saturday of the football season.

There is competition between species for the resources needed to sustain and continue life – there are other creatures eager to help themselves to the invertebrates in the meadow and the fruit that lie rotting on the ground. But there is also competition within species: between the individuals of that species, between the pairs of that species, between the families and sometimes the flocks and herds and swarms of that species. It's about gaining an edge. The Outlaw Josey Wales liked to approach an enemy with the sun behind him. The blackbirds in the meadows and beneath the apples trees are playing chicken with winter: staying as close to the breeding grounds as they can without actually dying. I could witness the struggle for existence every time I took a sip of tea.

Evening chores. Blackbirds sound farewell to a soft sweet day.

I suppose it's always changing, always reforming, never quite the same from one moment to the next, and no doubt that's part of the point of it all: what deep familiarity with a piece of landscape is really all about. But most of these transformation scenes are subtle and cumulative, hard to catch with our human perceptual equipment. You need a stop-frame camera to get the real drama of the transitions that come with weather and seasonal change.

When I talk about landscape, you must understand that I also mean skyscape. The two things are pretty well indistinguishable in this part of the world. You can see the sky all the time, even when your head is slightly lowered. A level gaze from a standing position will more or less fix your gaze on the horizon. I have stayed in more lumpy parts of our own country and been unable to see the sky from the bedroom window: green fills it from top to bottom – soft green grass that can only be grazed by cramponed cows able to rope themselves together or fitted with one pair of legs longer than the other.

Here on the Broads the clouds are as much a part of the landscape as the trees: looking out and looking up is pretty much the same thing. Sky, land, water: concepts that get fuzzy round the edges if you happen to live here.

It's also true that if you have outside tasks to perform as the daylight fades, the hard division between day and night will also get a little bit fuzzy. Like the barn owl, I am a crepuscular creature, at least at this time of year. It's a strange thing: you're in the house, and it looks pitch-black, but you step outside and within a few moments you're quite comfortable in the undecided grey light that comes before the night. The vast skies make for a long and generous twilight: that luminous, nacreous grey – grey has never been a colour much loved by romantics, so perhaps we'd better call it silver – is the prelude to every winter night: a twilight that's fifty shades of silver. Argent, you would call it, if you were using heraldic terms: you would begin your description of an escutcheon 'on a field argent . . .'

I brought the horses in from their field argent and gave them their food, never an act that goes unappreciated. Eddie loves to take part in this ritual, but that evening he was still on his way home from school. (By evening, I mean it was a little after four – the solstice was approaching, the afternoons short, the twilight coming earlier and earlier.) I switched off the stable lights and in the instant of darkness it was as if my own senses were instantly sharpened by the throwing of the same switch. Yes indeed. It was them. Them again. First I heard them, then I looked up and saw them.

On another field argent – the silver expanse of the sky – a thousand geese sable. In an instant of time the landscape or skyscape, insofar as they can be differentiated, was a different place altogether, the land of softly honking geese making the same glorious calligraphic swashes and flourishes they had inscribed over the marshes to the east when I had waited there with Jake the other night.

For a long minute I was alone with them all: how often, I wonder, can you be so far from other humans, in the company of so many non-human creatures – of such impressive size and volume – in lowland Britain?

There are moments of intense involvement in wildlife:

when the things that divide one species and another – the things that divide us humans from everything else that lives and breathes and transpires on this planet – seem thin. There is the Celtic idea of thin places: places where the boundary between earthly and heavenly things gets a little less solid than it does in normal everyday sort of places. Certainly I'd be inclined to say that this bit of marshy land is a thin place, for here the boundary between human and non-human life seem that little bit less solid, that little bit less permanent, that little bit less important than it does elsewhere. And even in such a place, there are times when the boundaries are thinner than others.

Goodnight geese, goodnight sweet geese, goodnight, goodnight.

Blackbirds

by Eddie Barnes

blackbirds
on the ground
in the garden
looking for food
in our garden
lots of apples
blackbirds feel better then
happy
they sing
to their friends
and tell them
where the food is
then the garden

looks black
with blackbirds
they all fly down
and feast
like Christmas
they should have blackbird crackers
to pull with their beaks
they could tell blackbird jokes
wear party hats
and run around
laughing

6
Transformations

Morning chores on a slightly longer day than yesterday. Dunnock sings a song of celebration.

'They even fug up the spectrum.'

A world-weary groan of protest from one of the two schoolboy heroes at the start of Julian Barnes's *Metroland*. The problem was the sodium lighting 'they' had installed in the cities. 'The colours. The street lamps. They fug up the colours after dark. Everything comes out brown, or orange.' They? 'The unidentified legislators, moralists, social luminaries and parents of outer suburbia.' There's a lot of 'they' about in the 21st century, perhaps in all centuries, and they do a lot of fugging up.

They even fug up the seasons.

It was the December solstice: the most southerly declination of the sun as the North Pole tilted away from the sun at an angle of 23.5 degrees, as if it were doing everything it could to shy away from the life-giving warmth of Sol. This is the day with the fewest hours of light: to be precise, 7 hours, 49 minutes and 41 seconds before 'they' have to turn the sodium lights back on again. That's nine hours' less daylight than at the June solstice. When you look at these figures you cease to wonder why we're all a bit mad. There is a touch of bipolarity – more or less literally – about all of us who live in such extremes of daylight.

I was feeding the horses before the sun rose: that is to say, before four minutes past eight. I would be giving them their evening meal only eight hours later, in a murky shade of silver. This was the darkest, most dismal day of the year: the bleak midwinter; the day the year hits rock bottom and then goes for the dead-cat bounce. A things-can-only-get-better feeling overwhelms us at this time of year, and that's what brings us our great winter festivals: the year has done the hard yards and, no matter how difficult it gets, it's going to be lighter when we struggle through the next day. More silver, less grey.

And that's all very well, but nine days earlier I heard a great tit singing. He was belting out the twin-syllabled song with which he welcomes the spring: strident, defiant, almost absurdly optimistic. And that's really why I hate the 'they' every bit as much as the unrelated Julian's clever schoolboys.

The first birdsong of spring should be a moment of unambiguous rejoicing, of realising that the world is fighting back: that the days of warmth and fulfilment are on their way back again – because, more than anything else in the world, the coming of spring is an unambiguously good thing.

But of course it isn't any more, is it? These days the rejoicing at the first sign of spring has been taken away from us. The ambiguities of climate change have left us struggling, no longer knowing the difference between good and bad as the seasons change around us. There is a kind of moral fog when it comes to these moments of transition. Ivan in *The Brothers Karamazov* said that if you don't believe in God then everything is permissible: there were no longer such things as good or evil. The muddle of climate change has created a similar loss of certainty: our hearts want to rejoice when a great tit sings out, establishing for himself a territory in which he will able to attract a mate and raise a nest full of more great tits – but we know that this incontinent hustling of spring into bleak midwinter is not the death of winter, as it was in times past, but the death of certainty.

It was school holidays, so Eddie was about at the early hour required to get the horses in from the field and able to lend a hand. Dominance hierarchy is a big thing in horses, and it's something a horseman always keeps in mind. When you turn the horses out in the morning, the first horse to go out is the last: the one at the bottom of the heap. That's so those higher up can't hang around at the gate and get in the way and get all hierarchical when you're bringing in a lesser member of the group. So Molly goes out first, Norah second, Mia last. Reverse that order in the evening, for the same good practical reasons. Eddie led Norah in, a horse who takes a little handling at the exciting time just before supper, and then he took Molly, with his usual lack of fuss. He was good at this, operating with a quiet, unforced self-confidence. When Norah jumped and half-turned, he stayed calm and steady and Norah stopped messing about: without speaking he had just told her there was no need for any nonsense. No raised voices, no threats: just good, confident body language.

Eddie had spent another day at Clinks Care Farm. This is a heaven-sent establishment a few miles away. It produces crops, and it has cows, chickens, sheep, goats, ducks, alpacas and pigs. Eddie goes there to work: and they work him all right; he comes home knackered. He does hard physical work in demanding weather. For a person with lax muscles and ligaments, this is not a straightforward business. Eddie loved it from the start.

The place offers the privilege of tough, meaningful work in the open and it's available to people with a great range of difficulties: mental-health problems, learning difficulties, physical problems. There are staff and there's a network of volunteers. Eddie told me he had spent most of day cleaning out a cattle trough because there was new stock arriving the following week. It was rather a business getting that trough right: a day of scrubbing. Next time he went he would have the reward of seeing cattle eating from the trough that he had cleaned himself: that he had changed from a sable silvered to shining argent.

Before Eddie came out to get the horses in I had done the mucking-out. It's never a task I have resented. It would be too much to say that I like doing it: but there is still a sense of privilege in doing the work: in filling the barrow full of droppings and wet, stinking shavings, in emptying the barrow on the muck-heap, in the sweeping and clearing and changing of water and adding the hay to the boxes. It's a serious privilege to have horses, and therefore the task of keeping them alive and healthy also has some kind of privilege attached to it. Meaningful work is a kind of privilege.

It was dark. Eddie checked the horses one last time – 'three happy horses!' – and we went back in. It was colder than it had been when the great tit was singing. Not forbiddingly cold, but we were glad to get in. The Christmas tree, vast as always, its tip bent at a right angle by the ceiling – it typical of Cindy's generous heart that she always found a tree just a fraction too large for the space – was shining gently all around. The Christmas cake had yet to be decorated, though. Eddie and I had made it a couple of weeks earlier, and quite a lot of the cherries got into the cake.

> Robin watches me at evening chores. Look, someone's got to get the stable ready for Christmas night.

When the great tit calls, you have to look at the marsh in a different way. Before you know what, the damn thing will start growing on you. Literally, I mean. The plants will take things into their own hands. Some will get bigger; others will appear in places where there was no visible sign of them before. I don't know – you do your best to be hospitable to plants and then they start taking the place over. There have been years when we've been unable to get out onto the marsh at all for several weeks because we missed the window for dealing with the plants. Had to wait for the autumn die-back to get back onto our own land. That's the problem with wild things: they're not terribly tame.

Space and time. Perhaps every book ever written – and certainly every novel – is about space and time: a little less than 24 hours in Dublin; 15 years across Russia; a century in Colombia. What we tend to forget, in our citified lives, is that space is no more passive than time. Space is an active character in any narrative: difficult, dangerous, often uncontrollable, and having its own will on the way the story goes.

Don't think I'm being fanciful. Not for a second. That marsh out there is jam-packed full of its own plans, and if we left it to make all its own decisions, it would probably – in the space of a few hundred years of solitude – become a closed-canopy oak forest.

That's the sort of thing landscape does. And you can say – as some proponents of rewilding do – that we should leave the land to do exactly as it wishes, and that to make any intervention at all is heretical: anti-life. Which is a fair point and a strong argument, but where will the herons and the egrets go fishing when the dykes have dried out in the oakwood? True, humans dug those dykes, but is that a reason to despise them? When I saw a little egret working the dyke, with the breeding plumes just starting to get a little extravagant – they won't be in their full glory for another few weeks yet – I was inclined to think that the dyke and this open, wet landscape was a fine thing.

I've always preferred egrets to abstract ideas.

What we're dealing with is the natural succession of vegetation. Left to itself, the marsh – our meadow and the big one next door, the wet grazing marshes of the common on the other side, the belts of maize on the farm, planted to feed and shelter pheasants, the fields of winter-sown wheat and sugar beets favoured by many of the farmers all around us, and Jake's hairy field margins and lofty hedges – will all join together and form that vast oak forest.

Take the common: grazed enthusiastically by sheep, intersected with dykes, covered in rushes, classic Broadlands landscape. It is, in its way, as artificial as Eddie's vegetable garden, where he and Cindy grow beans and other stuff, especially pumpkins, their speciality. If they stop looking after it, vigorous annual plants will set seed and out-compete the vegetables. The grass from the lawn will encroach. In the same way, out on the common, if the sheep stopped grazing, the grass would grow long and other plants would come in and start to grow up. And indeed, I'd love to see that happen, to see what plants released by the absence of grazing pressure, see what animal life is attracted to the changing acres.

If you were to leave the common – or, for that matter, the vegetable garden – untouched for several seasons, the tougher plants would colonise, then the brambles would come in and form tangles. Within these tangles, pioneer trees, like the sallows, would grow up. Eventually the trees would dominate, and as they dry up the soil and lay down humus – 'not taramasalata, then?' as a friend asked when I was making this explanation – the oaks would move in. And a closed-canopy oakwood is considered the climax vegetation of lowland Britain. (Though in an atavistic pre-human landscape, large wild grazing mammals would have created and maintained clearings.)

So if you don't intervene, the place will change. It will become a quite different place. It follows that doing nothing is a violent form of action. Working like a slave to keep a place as it is – intensive farming, in other words – is an equally violent assertion of the will. As is mowing the lawn: you're not just making the grass shorter, you're stopping the lawn turning into an oakwood.

So it was, then, that we did a small bit of management and got rid of the sallow blob in front of my window. Not just to give me a better view but also to keep the marsh marshy. We also cleared an area of marsh on the far side of the dyke from my hut. I wondered what this small act may reveal.

To act or not to act. When is it right to act? If ever? This is an eternal dilemma, a question without an answer, a problem without a solution. It has bothered us off and on from the moment we arrived and took possession of the first section of marsh. I have been lucky in having the advice of good friends who work professionally in conservation: notably Ian Robinson of the RSPB, and Julian Roughton and Dorothy Casey of Suffolk Wildlife Trust. All agreed that the place was special: all agreed that the specialness was at least partly because no one had touched the land for years. That's not only interesting, that's rare. There's always that urge to interfere.

They were mostly agreed that the marsh would change character – cease to be a marsh – if we allowed the willow scrub to take over. The trees would suck out the wetness and make the place into something else.

So we took out a patch of growing scrub. Our old friend Dave Barker came over with a chainsaw to do the job with Cindy; they stacked the stuff they took down – the brash – to create more habitat, shelter for small mammals. Neither was very comfortable with this. 'The thing about this place is that no one has interfered,' Dave said. 'And now we're interfering. It doesn't make sense.'

Cindy was equally troubled: 'The branches we took down had clearly been a shelter for the deer, and it doesn't feel right to take them away. But it also seems right to listen to experts.'

What feels right? What is right?

Stable chores on Christmas morning. Curlews merrily on high.

We would set off – never before two in the morning – to The Venturers, the all-night cafe in the Cumberland Basin in Bristol. The route took us through the docks and over the lock gates, towards a dredger that seemed permanently moored there, a great chain of buckets draped over its deck. As we passed it on our way to bacon sandwiches and pints of tea, we would make a ritual reference to an ancient joke: the dredger was in fact an eccentric millionaire's luxury yacht. It's probably unnecessary to add that we were students at that time.

My friend Tone often accompanied me on these late-night excursions. And he was reminded of it when I took him for a ritual walk around the marsh. It was a few days after Christmas, the tree still shining, the cake not quite eaten.

Tone, like many of my friends, is broadly sympathetic to the cause I have adopted, and the way I have given practical expression to it with these few acres of marsh. He gets it intellectually: but he doesn't really get it in his heart and soul and guts. (He has his own cause, which is saving the NHS.) So I took him out for a walk around the marsh and he couldn't have been more supportive. In fact, he understood the place perfectly in every way but the way I did. So much so that when we stopped for a while and drank tea from a flask, I saw the place a little as he did: just rather a mess. It's really not at its best in January: cold, windblown, silent. What's the point of it? It was then that Tone laughed and told me: 'It's like the eccentric millionaire's luxury yacht.' Meaning, I think, that it didn't look much to the average passer-by, but it meant a great deal to the owner. Who had got it right, then? The passer-by or the owner?

Another friend from those days pointed at a painting in a gallery and spat out the words: 'That is a great work of art – if you but knew it.'

Now I know how he felt.

Among my trove of excellent Christmas presents was a megalomaniacal torch: the sort that throws a beam thousands of yards, with several million candlepower behind it, so you can sweep the beam around like a lighthouse. I have used such powerful beams in Africa, looking for creatures of the night, and it is the most excellent sport. On a big night you can find leopards.

Less chance of that round these parts, but there is always scope for a person with a truly offensive torch. What you're looking for is eyeshine: the reflection of your light in the eyes of the mammal you've trapped in your beam. We primates mostly lack that reflective patch – the *tapetum lucidum* – because we tend to be exclusively daylight beings. But most of our fellow mammals have this device, which is a great aid to night vision.

So I swung my torch across the darkness and found two

jewels shining out of the black: a deer on the meadow, having slipped through the post-and-rail fencing that separates it from the marsh. I suppose I should have sworn at it for taking the grass from the horses' mouths, but I merely nodded and, after a moment, turned my beam away apologetically. I felt for a moment as if I had produced the deer from my hat.

Prizefight in the sky: cawing crow v battling buzzard. Must be Boxing Day.

Charlie Burrell, knight baronet, farmed the 3,500 acres of the Knepp Estate with the most colossal intensity for 20 years. And still failed to turn a profit. So he put a fence round it and let it do what it liked. The most violent form of action, in short.

The details of the story are complex and there were any amount of official problems in the setting-up and the maintenance of this gloriously demented project, but that's the essential truth of it. He threw his hands in the air and told the place to get on with farming itself. (You can read the full story in Isabella Tree's excellent *Wilding*.)

The idea was to recreate the ancient fauna of this country, to see what they did to the ancient flora of the place. He put in free-ranging English Longhorn cattle to play the part of the extinct auroch or wild ox. Instead of wild boar – you can't have free-ranging wild boars; it's against the law – he put in Tamworth pigs. He also introduced ponies and fallow deer.

'The whole thing is process-led,' Charlie explained when I made a memorable visit there. I asked him to elaborate, and he did so, but I still wasn't quite clear. Then I thought I had it.

'You mean you don't know what the fuck you're doing?'

'Precisely!'

And that spirit of boldness has seen a number of extraordinary and unexpected things happen on this land, including a population of nesting nightingales in a habitat that was previously considered useless for them; huge numbers of purple emperor butterflies, previously considered exclusive to mature woodland; and a glorious invasion of hideously endangered turtle doves, again in circumstances not recommended by traditional conservationists.

So what about that closed-canopy oak forest? Plenty of oaks around on his land, but no sign as yet of a takeover. That's because the large grazing mammals keep parts of the ground open, while the pigs turn over the soil when digging for roots. They change the dynamics of the habitat as surely as a plough and flail.

So what's natural and what's not?

Out in the darkness the Chinese water deer got on with their job of managing the habitat. Whatever that is . . .

> Morning chores in a mist of sterling silver. That slightly less dense patch is a barn owl.

I have known Ian Robinson since he was smashing up reedbeds for a living, so it's always good to see him when he drops round, even if he does look at our own reeds with a slightly dangerous expression on his face. He was the great pioneer of reedbed destruction, and it's a title he rightly carries with some pride. Here was intervention at its most drastic, and it certainly didn't meet with universal approval. Not at first, anyway.

He was number three at Minsmere, the great RSPB reserve

on the Suffolk Coast, at the time. That was 1990, and research had indicated that the great reedbeds of Minsmere – the reserve's signature habitat – were no longer suitable for bitterns. The birds were continuing to decline fast, despite all kinds of concentrated conservation efforts.

Bitterns are seldom-seen birds that lurk below the level of the seed heads in the reeds; in the spring the males announce their presence not with song but with a glorious far-carrying hoot that is, despite the wrens' best efforts, the loudest bird noise in Britain. The conventional thinking in conservation was that if you made sure there was a decent expanse of reedbeds, you would get plenty of bitterns. But apparently not. Even at Minsmere, the birds were declining.

It wasn't just about the acreage of reeds, the researchers declared. It was also about their quality. The bitterns needed young, fresh, wet, vibrant reedbeds. The stuff at Minsmere was too old, too dry, and no longer held the life that bitterns needed to feed on. The reedbeds had been intensely managed, and the encroachment of scrub and bramble was routinely under check at times of the year when the bitterns weren't trying to breed. There was no danger of the reedbeds becoming an oakwood . . . but all the same, they were too dry. They were old and tired. The suggested solution was radical: dig the beloved and precious reedbeds right out. Destroy them so they could start again with their feet in water.

Ian was in charge of delivering that. It was a heartbreaking business that involved extensive trashing of reedbeds. Not everyone was convinced that this was a good idea. It looked exactly like the mad, wilful destruction that the RSPB was there to try and prevent.

Ian doesn't do half-measures. His life has shown a pattern

of complete devotion to the cause of wildlife conservation. And as the young, green, new reedbeds began to rise up in the places he had reduced to mud, it was the beginning of vindication. It turned out that the research was right: and the bitterns multiplied and thrived. The great comeback had begun. Since then Ian's work has been replicated at reedbeds around the country.

Ian is now in charge of the RSPB's work in the Broads. Before he got to know the place, he was worried that his job would be about navigation and dredging and the rights of pleasure boats over everything else that moves. Like me, he became a convert. This is a landscape he understands and longs to do more with.

We walked round the marsh, the brief snow now quite gone, and he recommended that we dig a scrape: a wide shallow pool. This was a thrilling suggestion. Such a feature would bring in waterbirds: it would be a feeding-station for those passing through, and a major resource for any birds that chose to hang around and breed. The most famous scrape is at Minsmere, where Ian worked for all those years. It's the best place to see avocets, the bird on the RSPB logo: they went extinct as breeding birds in this country, but came back after the Second World War. The land on the Suffolk coast was flooded to make invasion from Holland more difficult and, quite by chance, this created the perfect conditions for avocets – so it was an air force of avocets, rather than the German army, that did the invading. Ian was convinced that a scrape on our bit of marsh would work, and he had great experience.

I always get the feeling that Ian would love to spend a day trashing our place. In the nicest possible way. One of the things about being a practical conservationist is that you get to see land as a project. You are constantly aware of the way land changes over time, and you can always see something different in the land before you. Conservationists are used to looking at land and seeing how it would look in five, ten, 30 years' time, or even a century or two. It was something I was beginning to learn myself.

Ian would like to take out a great deal of the scrub and the rank grasses. In some moods he would love to enclose the place securely for livestock and put a herd of, say, Highland cattle – a great favourite of conservationists – onto the marsh for the autumn to eat the place down. He looks at this scruffy, scrubby patch and sees a field full of orchids. And while I have plenty of time for orchids – we have had southern marsh orchids in fair numbers out there – they're not my priority.

And managing land is always about priorities. That's true if you are growing sugar beet down the lane or growing bitterns at Minsmere. That's true if you are growing geraniums in a window box or pumpkins by the house.

Morning ride. Is that's what's left of the sparrowhawk's dinner?

A week or so later, the place was transfigured. We had a good chunky fall of snow – one of those classic transformation scenes: hair down, glasses off, my God, you're beautiful! Perhaps you remember the scene in *Pretty Woman* when Julia Roberts meets Richard Gere after she's been to the posh shop in Rodeo Drive. She's looking a million dollars . . . and yet just about everyone who ever watched the film – every heterosexual male, anyway – says, well, she was just as gorgeous – even more gorgeous – when she was wearing the hot pants and the boots with a safety-pin zip. Transformation is a wonderful thing, but often it reveals what was there all along – something you were, for some reason, unable to see. It's not that the beauty is new-formed: it was always there – if you but knew it.

All so very white, the sky a rather ridiculous shade of blue, and out of sight, from the water on the far side of the river, the sound of greylag geese. It wasn't the thickest fall of all time, but enough to give the ground an uneven coating, just as Eddie managed with the icing on the Christmas cake. Deep and crisp, uneven. There was just about enough snow for Joseph to pelt Eddie with snowballs; enough for the two of them to make a tiny snowman that would have been only slightly too big for the cake.

And out on the marsh, time too was frozen. Here in the sudden change was a record of time past: by walking out today I could see what happened last night: here passed deer, and here, an otter, the webbing between its five toes clearly visible in one especially clear footprint. In this long, fragile moment before the weather changed again, each seed head of the reeds bore a few crystals of snow, precariously balanced, the sun shining through it in a beauty as self-conscious as Julia herself in her Rodeo Drive finery. The black branches of the sallows wore white shirts of impossible cleanness.

Tone's notion of the marsh as the eccentric millionaire's luxury yacht was about gratification of a non-obvious kind: beauty hidden to all except those who did the owning and the managing. Under this fall of snow, it was as if the dredger had been hung with a million fairy lights, its beauty suddenly and rather provokingly obvious to all. The marsh, bracing and untidy and scrubby, had been transformed into this exquisite Sisley landscape or perhaps into that Parisian snow scene painted by – you'll know the picture even if you've forgotten the name of the artist – Norbert Goeneutte. So there was the marsh, decked out as the Boulevard de Clichy under snow. It has shifted to a charming but hackneyed beauty: the Boulevard de Cliché.

But no wild beauty is truly hackneyed: it's all in our own reading of it. Behind the beauty lay a killing cold: and all around me – and, for that matter, below – there were creatures struggling to survive. Somewhere out in that snow there were peacock butterflies, passing the winter in diapause, waiting for the time to take wing in multicoloured, many-eyed glory. The true beauty of the snowfall was the pause between the falling of the snow and the rising of the peacock.

Morning walk for me means a morning gallop for a hare.

Mr Wright's driver arrived with his digger to perform his act of transformation. How strange our bit of marsh looked with that great monster of machinery turned loose on it. Earth-moving machines always look like dinosaurs: here was a Jurassic marsh with Triceratops – or some other beloved beast from my boyhood favourite, *The Golden Treasury of Natural History* – roaring his defiance at time and space. It would surely be the dinosaur's world forever.

When we first had the marsh we were worried that the dykes were getting clogged up. The water – sluggish at best – had in many places stopped flowing altogether, and as more and more stuff grew up, we were worried we would lose them: no longer waterways but long, damp patches of scrub, scrub that was all the time drying out. What should we do? The costs of hiring heavy machinery and the people to operate it were frightening. When Mr Wright first turned up, it was a miracle cure for anxiety.

It was also a rather odd feeling: we were not alone. The way we manage our land affects other people. It's important that the dykes are kept open. It's the landowner's own choice, but the fact that the dykes can be kept open without trouble and at minimal expense stresses the fact that what you do with your own piece of land is not a personal and private decision. It affects our neighbours. And, in a larger context, it's easy to see that the way you manage any piece of land has implications for the surrounding area – for the country – for the planet. Example: if you concrete your front garden in a suburban street you add to the problems of flooding because you have taken away a place where heavy rain can soak away. Every decision you make about land is ultimately the business of everyone else on earth.

The digger-driver was perfectly willing to take on a little additional work for an appropriate sum, and we showed him where the scrape should be: basically an extension of the pond that Barry had dug a few years before. Not too deep, and not too uniform a shape, and long shallow edges: a scrape has become a conservation classic.

He knew all about this, of course. So we asked him: could you please make us a scrape? He was perfectly willing to do this: he just told us it would never work, we were wasting our money, the water would seep away, the peaty soil would never hold it, we'd just be making a big shallow hole. Mud was all we'd make: no water would stay and no watery bird would come.

He left us to consider this while he went about the job of clearing the dykes. It's always instructive to watch people whose skills have gone so deep they no longer require input from the conscious mind, giving the false impression that the task is easy. There was no swagger, but the absence of conscious striving added a wonderful illusion of nonchalance. I've noticed this same thing when very close to top-class athletes at practice, with Joseph playing arpeggios and scales, with Cindy shaping a piece of wood that will become a tiger or an otter. Here the digger transformed the dykes into frank, untangled waterways, stopping them in their tracks as they were in the process of taking the early steps towards becoming an oakwood.

Cindy, persuasive in most areas of life, was unable to inspire the driver with belief in the viability of the scrape. She reasoned rightly that there was no point in cajoling a craftsman into a job he believed doomed to failure. However, he agreed to clear the pond of reeds: so my idea of a reedbed for a possible reed warbler went back to open water. I hoped that was the right decision, but even if it wasn't, the reeds would come crowding back as fast as they could.

There is always something rather fine about inspecting the marsh after such a transformation has taken place. The day had shifted gear on us and become one of sudden warmth. As I walked along the bare earth of the dyke, I could hear great tits, blue tits and dunnocks in song. Two buzzards flew overhead, also two herons: there was a warm, thrilling, pairy-uppy vibe beginning to spread across the landscape.

Two great tits fizzed across one of the open dykes, one in full pursuit of the other. I couldn't tell whether this burst of action was punitive or amatory – and that's one of the great ambiguities of spring. Love songs and war songs sound the same, and every pursuit is equally likely to end in copulation or chastisement. Only the birds themselves know the difference.

I sat by one of these new-shaven dykes, the mud soft and fresh and wet, a thrilling smell of potential growth, the vegetation shoved aside by adroit gestures of the digger bucket. A little egret dropped down and began to feed. There is always something slightly incongruous about a bird as exquisite as an egret feeding with such enthusiasm: it's like a stained-glass window of an angel eating a Marmite sandwich. It's a classic clash between human fantasies about birds and the birds themselves. They are out there seeking life, no matter what symbolic load we heap on their shoulders. Perhaps it's the tension between these two states of being – birds as they are, and birds as creatures of the human imagination – that gives birds their special fascination for humankind. Certainly I enjoy egrets as angels just as much as I enjoy them as a small species of heron.

There was still an hour to go before dusk, when another angel appeared: another white-winged being that seemed there to bless the marsh and the transformation it had undergone that day.

Here was a barn own, exploiting his power of silence. The sun was still in the sky, though sinking fast, but it wasn't the great floodlight that illuminates the entire world. It seemed to be the narrowest and smallest kind of spotlight, one that existed only to pick out the white wings of the owl. The bird sank to the ground, out of sight for a short while, and then rose again, continuing its slow, silent inspection of the digger's labours. It dropped again, clearly finding great profit in all this disruption.

There was more than a hint of spring in the air, and in a fragile world there was promise of renewal: war songs and love songs, feeding frenzies, sex and new life. All across the marsh there was the earth and the water that would fuel that process.

I had a sudden memory of that line in *The Rainbow,* which was a great favourite of Ralph's when we were both sixth-formers arguing about books and the meaning of life. 'But heaven and earth was teeming around them, and how should this cease?'

These days we could all come up with a dozen reasons why earth should stop teeming; we might in certain moods argue that teeming was already a thing of the past. Everything we love is vulnerable, we all know that. But in these modest few acres there was a certain amount of teeming going on.

7

Being Magnificent . . . Nudge-nudge, Wink-wink

Morning chores and a squadron of goldfinches. Charm offensive?

Birds of prey were once cherished above all others. A raptor is always top bird: when tamed, the bird of prey became an expression of rank and fortune – the power of the noble owner. Falconry was the pastime of choice among the rich and powerful; the flown hawk was a living symbol of its owner: a great soul taking wing, the terror of the world. A falcon was *the* noble bird.

There's another reason why humans had such strong feelings for their captive birds of prey. All birds of prey have their eyes on the front of their heads, facing forward: what matters to them, in terms of perception, is the narrow arc of stereoscopic vision in which they can appreciate the world in three dimensions. That's how they make their hyper-accurate judgements of speed and distance.

We humans also have forward-facing eyes and two-eyed vision. Our ancestors needed to make equally accurate calculations about speed and distance, but not to catch birds. They needed to move with speed and safety from branch to branch. All the same, we can see something of ourselves in the face of a hawk: and the two-eyed glare we get back is a comparison that flatters us at a very deep level.

And here is one more reason for the closeness of humans and hawks: when you carry a hawk on your left wrist (I have done this on a few occasions) the hooked bill is of necessity very close to your own left eye. You can't hold the hawk at arm's length: that would soon tire you out. You must tuck your elbow into your side: and that puts beak and human eye just a few inches apart.

In other words, this is about the immense privilege of trust. Trusting, and being trusted. The extension of trust beyond the species barrier is a powerful experience for us humans. We seek it by bringing carnivores into our homes, onto our laps, onto our beds. We keep horses, who could lay us out with single irritated movement, and we even get on their backs. And the ancient arts of falconry require humans to put their trust in a bird cherished for its ferocity. (I am left-eye dominant: my right eye is a second-rate performer. To accept the proximity of that beak to my favourite eye was a powerful experience.)

But then we turned against them. We invented the shotgun and declared war on all birds of prey. A hooked beak was no longer to be tolerated. A bird of prey on your land was clear evidence of failure. At a stroke, birds of prey changed from beloved companions to hated rivals.

During the great Victorian persecution, the numbers of birds of prey fell dramatically. Five species were down to fewer than 100 pairs: golden eagle, hobby, hen harrier, red kite and Montagu's harrier. Five species were driven to extinction in this country: goshawk, honey buzzard, osprey, white-tailed eagle . . . and marsh harrier.

I could see three of them. All of them over – and indeed on – the marsh. Marsh harriers. All of them female. As I've been saying, some conservationists would look at the marsh and say: you could do better; I wouldn't manage it like this, I'd manage it like that. But here were three marsh harriers and every nuance of their body language said: you're not doing so badly; this place isn't so shabby, it'll do us. And what's more, I was able to keep all three of them constantly in view: the removal of the sallow blob had made that possible. The marsh was more open, more to their liking: and as a bonus I was able to gourmandise on the sight.

One of them dropped down to the ground. My guess was that she'd caught something and was devouring it at her leisure, while the other two were inspecting the place without getting too close to the female on the ground. This was, for all of them, an exciting encounter at an exciting time of year. There may not be highland cattle down there, and perhaps there won't be so many orchids, but there is stuff that keeps a marsh harrier going. This was a place worth inspecting at close quarters.

Marsh harriers started to come back to Britain after the Second World War, though in 1971 – after all birds of prey had been hammered by the use of pesticides – they were back down to a single breeding pair. That was at Minsmere, Ian's old place. By then, they were protected by law. What's more, the reedbeds and marshes they love to hunt over were now places cherished and maintained by conservation organisations. Marsh harriers started to spread again, doing so throughout the wet and watery landscapes on the eastern side of England. They're now at their highest levels for 100 years: in 2005 there were 375 breeding females. (Marsh harriers are counted by numbers of breeding females, rather than pairs, because some males will have two or even three females.) What would Charles Rothschild have made of that?

It was round about then that a strange thing happened. Some of them stopped migrating. They began to overwinter in England. It had been accepted that all marsh harriers went south for the winter. Now they were staying on: and at Hickling Broad, not a million miles from my place, I have seen 50 in the air at the same time in January. In winter they go there to roost together, before separating in the morning to set out and forage.

These magnificent birds—

—don't you hate that? All birds of prey are magnificent, apparently. So is every stadium that ever held a major sporting event. There seems to be an adjectival paralysis that seizes people where birds of prey and stadiums are concerned.

I have never worked out why it's so important for birds of prey to be magnificent, nor why their magnificence is relevant to their conservation. Is it more important to conserve magnificent birds than humble and homely ones? Is their impact on human emotions a relevant factor to the urgency of their conservation?

Perhaps what matters here is the top-down concept of conservation. There are many plants out there on the marsh, many invertebrates, and as a result good numbers of small birds and mammals. And because of that, the female marsh harrier was able to drop in and feed. The magnificence or otherwise of the three visiting marsh harriers shows that things are in half-decent shape: marsh harriers on the wing, all's right with the marsh. Even if the vole or whatever it was being consumed might regret his own role in that economy.

So the marsh is not *for* marsh harriers. It's for everything. But if there's a marsh harrier making a living there . . . well, everything – everything else – must be doing all right.

Morning ride. How dare the egret claim to be more beautiful than my horse?

Let me try and make you see a marsh harrier, for – spoiler alert – this is not the last page they will turn up on in this chronicle. They are birds of prey, but not of the quick and dashing sort. They operate at the opposite end of the speed spectrum. A marsh harrier's power – Eddie, are you listening? – is slowness. It has a very low stalling speed. It flies like an aeroplane with its flaps down: it can fly safely at a slower speed than most other birds; an eagle would fall out of the sky if it tried to match a marsh harrier for slowness. That's air speed, of course: the speed of the air passing over the wings, not speed over the ground. This slow flight is a different form of agility, a different sort of flying skill: but it is a very considerable one and it is the way a marsh harrier makes its living.

It holds its wings in a shallow vee – a dihedral – and that gives it control during its wavering glides. They patrol the marsh, with a particular fancy for the edges of the dykes, and from their drastically gentle speed, they have the leisure to see everything that lives and moves underneath, and then to drop down like a shuttlecock onto anything that might look like a meal. Nothing dramatic about the pounce of a marsh harrier: it dips in the air – is gone. And if it doesn't get up at once, it's probably caught something.

Three females at the same time: that was deeply pleasing. They were no longer hunting: this was now an examination of each other and of each other's claim on the patch of country below. There was a sizing-up going on: no active aggression but a general eyeing-up. It seemed that this bit of marsh was a valuable property. It mattered: to those three marsh harriers, it mattered very much.

The females are dark brown with creamy heads. Their lines tend to be long and slim in straight and level flight. You can, with an idle glance, mistake them for a buzzard when they are soaring – gaining height without flapping their wings – but when they un-fan their tails and drop back into the dihedral, they are harriers through and through.

These magnificent birds of prey have become rather homely birds of prey since I moved in here: and my daily life is all the richer because of that. To live on terms of familiarity with a once-extinct bird is a good feeling. I remember being shown marsh harriers at Minsmere by John Denny, an old boy who used to arrive at the reserve at opening time every day on his tricycle and spend the day in Island Mere Hide looking at marsh harriers. When he saw one he would bellow at the top of his voice: 'Marsh harrier! Flying right!' I don't know if it was the birds' rarity or their magnificence that got to him: but to Denny this was the bird of birds. After his death he donated his copy of *Birds of the Western Palaearctic* to Minsmere's library, a highly acceptable bequest. It comes in nine volumes – I have them all on my own shelves, naturally – that cover, in considerable detail, every species routinely found in the biogeographical area in the title. Eight of Denny's volumes had scarcely been opened. In the ninth only one species account had been seriously perused: and that one again and again and again.

Birds of prey will do that to humans more than any other kind.

As the year advanced, so did the vegetation – reminding us of one of our more disastrous and costly episodes in management. It's not enough just to cut down vegetation: you need to clear it as well. If you don't, you're adding nutrients to the soil, creating an over-rich and unnatural environment. If a farmer maintains wide field margins, the job is not done with an annual cut. Jake at Raveningham also bales what's cut, and then removes the bales. That's why his field margins are bright with flowers and buzzing with bees.

We had tried creating the same effect by strimming and raking, but that was either time-consuming or expensive. So after taking good and proper advice, we bought an impressive machine that cut vegetation and retained what it cut. The idea was to make a pile of the cuttings on the marsh, which would make a nice warm rotting place where snakes could lay their eggs. Alas, the tractor was a complete dud. We spent a good deal of money trying to fix it, but we had to get rid of it at a considerable loss – and in the meantime the management jobs stacked up. Doing the right thing is never easy.

> Deer footprints look like inverted commas, but don't quote me.

The tiny common that lies to the north-west of us – about ten acres, so a little bigger than our bit of marsh – is grazed by sheep. Jane, the shepherd, is good at her job. It's a difficult business: farming people will tell you that, from the moment of birth, a sheep has one ambition and that is to die. Get even slightly behind in the system of prevention and they will be eaten alive by maggots. Fly-strike it's called, and it's a highly distressing business to clean up.

The sheep are good custodians of the land – being light-footed, they don't compact it as much as cattle or horses – and the place looks good. In certain moods I longed to get my hands on the place and manage it with wildlife as a priority. It's a classic grazing marsh: a Broadland meadow, in short.

A couple of miles down the road, and not entirely by coincidence, there is a place called Broadland Meadows. But there is no meadow there. It's a housing development; and I'm not arguing about the need for housing. It sounds lovely: fancy living at number 15, Broadland Meadows. But the meadows were destroyed to build the houses. The houses stand packed together, without so much as a garden, not an ounce of generosity in the planning. It's loudly agreed that we need houses; by choosing such a name, it's tacitly agreed that we also want – need – meadows.

The pattern of naming a place for what has been destroyed is found all over the country: names that might have been borrowed from Thomas Hardy are used for light industrial estates, motorway service stations and housing developments. And somehow everybody is fooled. By choosing a sweet especial rural name, we can pretend that we're having our cake and eating it: living a rural idyll, at peace with the nature, while having all the convenience of suburbia.

Little egret recalls Isaiah. How beautiful are the feet!

The end of January is the time for the RSPB's annual Big Garden Birdwatch. The best thing about taking part is that you must sit for a full hour, doing nothing but looking for wildlife. This is a large-scale, citizen science project: the previous year more than half a million people sent back their forms and recorded more than 40 million birds.

Naturally, my pride in the marsh means that I long to send our own form back full of impossible exoticisms: bitterns, cranes and Montagu's harriers, say. But this is about gardens, so Cindy, Eddie and I took a seat looking out at the bird-feeders and the lawn leading down towards the dykes and the marsh. That's not the country for bitterns or cranes. And anyway, you won't see a Monty in the winter; they have stuck to their migratory habits, the few that ever turn up in this country.

So there we were, at the dining table, with cups of tea and pencils and binoculars: and the usual sort of garden birds came to the feeders. That's all good fun: great tits and blue tits hanging upside down; blackbirds and dunnocks at the bottom; robins making fleeting visits for the softer food; pheasants walking up to scavenge on the ground beneath.

Eddie could identify most of these. The watch began, as always, in a mild frenzy of activity, and then, once the obvious birds had made their mark, the exercise became more contemplative. Under the rules of the Birdwatch, you're not allowed to count flyovers, which is bad for showing off. We couldn't count the little egret, which seemed unfair. However, I managed to pick out a female kestrel, perched in the willows at the far end of the garden.

We had already had a great shared wildlife moment in January, Eddie and I. It was a blue whale: not, alas, a living one, but the great model that lies in the Mammal Hall at the Natural History Museum in South Kensington: a wondrous thing I gazed on as a child, and which helped to shape my world. Eddie had grown fascinated by blue whales. Two different geniuses of wildlife had inspired this love. The first was David Attenborough: Attenborough's cry of joy – 'The Blue Whale!' – had lifted his heart: all his being was out there in the open boat in the middle of the ocean with the great hero: and he shared his joy as the greatest creature that ever lived came to the surface to breathe.

The second genius was Nicola Davies, who has written many wildlife books for children, including *Bat Loves the Night*, already mentioned in these pages. Her *Great Big Blue Whale* is a masterpiece – poetry and science in glorious combination – and I have read it to Eddie more times than I can easily guess.

I told him nothing about where we were heading as we set off from my father's place in London. He recognised the building from the film *Paddington*, and that was a good start. Then, as we filed in, we dodged the rush to the dinosaurs and went straight to the Mammal Hall: and there, 100 feet long and swimming through the air above our heads . . . the blue whale!

What could be better than that? Eddie held up his little finger to indicate the size of the krill, the blue whale's chosen prey. A still better moment came after we had climbed to the gallery and were walking around, face to face with skeletons of other cetaceans that were suspended from the ceiling high above. And as we looked at the extent of a grey whale skeleton – the famous migrant that makes biennial journeys between the Arctic and the Pacific coast of Mexico – Eddie called out with joy and reverence: 'Baleen plates!'

He had seen pictures of these mysterious things, he had seen Attenborough demonstrate their function, he had read and been read Nicola's description. Now he was confronted by the real thing: real baleen that had been used to sieve real food from water, the secret of the great whales' existence. So once again I told him of my own encounter with grey whales in Mexico, and how they came up to the boat to be patted.

Now at the dining table, we were making a list of these homely, non-exotic, decidedly non-magnificent birds, and spending a full hour at the task. The best bits came when the long-tailed tits dropped down. These are birds that love togetherness. They call to each other the whole time: the answering si-si-si to their own si-si-si is more than mere reassurance, it is ocular proof that all's right with the world at that particular moment. They drop down with their impossible shapes (fat little round ball, ridiculously long stick of a tail) and colours (pale and dark grey picked out with the most tasteful possible pink, the sort of thing they go in for in the lounges of four-star hotels) as they filled the window with their busyness, their pleasure in being together, and in the abundance of the food resource that they could find in front of our windows. But the most constant birds are blue tits, dropping in again and again to eat their acrobatic meals. Blue tits, blue whales: both equally magnificent, in their different ways.

The hour was up. Time for a late lunch. It was Sunday: when it was dark I would make a feast with four different curries.

> Morning chores and a solo from the red-headed drummer. Is that the great Ginger Baker or a great spotted woodpecker?

A couple of days later, January was over. As I walked the marsh, the pheasants gloated before my idle steps: they'd got away with it, then. As the year advances, so more and more of the pheasants scared by the beaters make a sharp right turn the instant they are airborne. As the shooting season develops they get better and better at it. They go hammering at right angles to the guns, heading straight for the marsh. Is this a learned response? Either way, the last shoot of the year on the farmland behind us is usually a brief affair. By the time it's finished it must be standing room only down on the marsh. The pheasant-shooting season ends on the last day of January: so now the surviving pheasants had the world to themselves again.

It's the same over on the flood. When I hear the whistling of wigeon in February I can allow myself a small smile. These are wild birds that the wildfowlers failed to get, for their shooting season is also over. The people who shoot the flood on the far side of the river are unappeasable: there is scant pleasure in hearing the whistling of wigeon during the shooting season, because you know their likely fate. But in February I can give a nod of congratulation to the ones that got away.

There were now eight months of peace available to pheasants and partridges and ducks and geese. And me . . . well, I've always preferred life without the sound of gunfire. But it seems that if you choose to live in a place full of nature – full of life – you find yourself living in a place full of death. Some of it is part of the natural cycle. Not all.

Chores on a frosty morning. The stock dove is amazed at the world's whiteness. Coo-er! Coo-er!

Today we were at the still point of the turning world. Like all other days, really. But on some days the stillness and the turning are more obvious than others. Sometimes the turning of winter into spring seems like a motorist trying to change direction in a lane only a fraction wider than his vehicle is long. To get out facing the right direction he must go forwards and backwards again and again: the multi-point turn they don't teach you at driving school but which all drivers have to master or at least bluff their way through at some point.

This was still unambiguously winter, but with sledge-hammer hints of spring thrown in. That's a chaffinch in full song, and his song goes: 'Nudge nudge, know what I mean, say no more!' That's the ancient Monty Python sketch of the man in the pub asking his neighbour if his wife is a sport, if she – know what I mean? – *goes*. Looking out at the marsh as the year starts to get on with its multi-point turn is very much like that sketch. It's as if someone is asking the marsh: do you *go*? Do you do ... *life*? Know what I mean? Are those chaffinches going to make lots more chaffinches? Are those plants going to sprout and flower and fruit and seed? I bet they are – I bet they are! Say ... no ... more!

And so I heard a song thrush singing out loud and clear. It didn't sing the following day, but it was a start, and spring is all about starts. Every advance is followed by a retreat, but the retreat is never quite as long as the advance. When a song thrush sings, it fills the valley: a phrase repeated two or three times, and then discarded in favour of another.

Another day, from the wood beyond the marsh on the right-hand – south-east – side, antiphonal drum solos from two great spotted woodpeckers: perhaps a territorial dispute, perhaps no more than a discussion about precise demarcation. But these two Ginger Bakers were each staking a claim to be the best drummer in the wood. Great spotted woodpeckers will thwock into a branch to excavate the invertebrates that live in the wood: but they will also make the violent far-carrying drumming, not for foraging but to make a noise. It's the same for them as singing is for a song thrush: but they are percussionists rather than vocalists. They choose a good resonant dead branch for maximum carrying-power. I have seen them – and heard them – doing so on the metal fittings of telegraph poles: a fiendish din that must frighten every rival for miles.

A flock of fieldfares, about 20 of them, was heading north, waving black tails in farewell. These winter thrushes – that's how we think of them in this country – were now heading to the spring-lands: to Scandinavia and northern Europe. With their triple quacks they had brought us winter – or had perhaps softened the blow of winter – but now their blood was stirring within them and it was time to start thinking about making more fieldfares. The winter thrushes were turning into spring thrushes.

And where were all those cock blackbirds? Where had the shiny black apple-eaters gone? Gone to hen-birds, every one – or most of them. Our resident birds had stayed behind, but the great gatherings beneath the apple trees or out on the Broadland meadow were now part of the past, and no doubt of the future. For now they had gone: seeking the spring.

I also heard a mistle thrush: and that's a song I have special love for. It's always seemed to me especially wild: as if calling the world to disorder. With the first mistle thrush of the year, I always feel that something has been accomplished: that the early skirmishes have been won; that it's time to regroup and make a proper assault on winter's fastness.

And to put some life back into the land: get that marsh humming and buzzing and tingling and vibrating and copulating again. Nudge-nudge, wink-wink – say no more!

Morning walk and a song thrush. Hate to repeat myself but song thrush! Song thrush! Song thrush!

No, it wasn't a marsh harrier. It was too slim, a little too elegant – though of course equally magnificent – and when it turned away, I saw that it had a white patch high on the tail. It was a hen harrier: to be more accurate, a ringtail. That is to say, either a young male or a female. Males that are not up to taking on a grown-up, experienced male and holding a territory of their own make this stance quite clear by looking like females. No threat, no rivalry: I come in peace.

The ringtails can look pretty similar – at least from a bad view – to female marsh harriers, and they fly in much the same way: the same dihedral, the same mastery of low-speed flight. They can turn up anywhere in winter, and quite often overlap with marsh harriers. But the two species are divided in their choice of breeding lands: where the marsh harriers prefer the wet country their name suggests, hen harriers move onto the uplands. Not marsh harriers but harriers of the hill.

Once they reach these uplands they often find trouble. They have been relentlessly, obsessively and illegally killed by the grouse-shooting industry. I have had my share of grief from making what has been called a 'controversial stand' in favour of the hen harriers: the controversy revolves around the question of whether or not rich people are subject to the same laws as the rest of us. Let me just point out that, according to Defra, there is space for 300 pairs of hen harriers in England; the previous year there had been three.

So I watched the ringtail make his way across the marsh and smiled a little grimly in his direction. It was like hearing the wigeons in early January. Good luck, bird. You're going to need it.

Morning chores. Three curlews piping overhead . . . flying in the direction of spring.

If you tune into the wild world in the 21st century, you're going to take on anxiety. I had heard and seen herons making for the wood on the left-hand – north-west – side of the marsh, just beyond our boundary. I had heard the early sounds of herons establishing their presence there: for the wood holds a thriving heronry. I had heaved a sigh of relief when the shooting season ended – and as soon as it did, they started shooting up the heronry. For pigeons: some birds never go out of season. They shot hard for two successive Saturdays and I was left – not for the first time – wondering if this year they'd done it. Would the herons really be up to a return after their places of resort had been so uncompromisingly shot?

The fragility, the vulnerability of the wild world is inescapable. There are times when I can savour the rural idyll of living by this stretch of marsh: but there are plenty of other times when anxiety and worse overwhelm me.

So let's get something clear. The marsh's function is not to give me pleasure, even though it frequently does. Its success is not to be measured by counting the number of good moments I have had there, counter-balanced by the moments of anxiety and disappointment. The marsh's function is to live: to bring forth more life: to feed and shelter wild creatures and to allow them to survive and to breed and to become ancestors.

It's like sport, and that's a subject I have spent a long time thinking about. Sport is not entertainment. The job of the athlete is not to give me a good time: it is to win in the most expeditious way possible. David Gower or Johan Cruyff weren't trying to bring pleasure to me or to anyone else when they did their finest work. They were trying to win. I could find my heart's desire or find despair in what they did: it was all one to them.

So when I look out on the marsh, I do so in the way that I watch sport. I accept that it's not about my gratification. Rather it's about the victories and defeats that take place out there. The ringtail, passing through, was a fine sight: perhaps he went on to found a dynasty, perhaps he was shot the following week. The shortage of herons made me fret: well then, fret. The marsh harriers were filling the marsh not only with themselves but also with promise: nothing wrong in enjoying that, but my enjoyment is not the point. It's a bonus, an acceptable bonus: but it's not about me. It's about marsh harriers.

> Evening walk and a ghostly shadow vanishing in the churchyard. The deer departed . . .

Then, for the first time in many months, there was a male marsh harrier harrying the marsh. Apart from that dihedral, you might wonder if male and female are the same species. The male is slighter than the female, smaller by a third, and as a result the flight is more buoyant and airy. And then, as he turns and catches the light, he is glowing in three colours: a warm chestnut picked out with the heraldic colours of sable and argent.

Such ease on the wing, such mastery of the air: and I was lost in wonder because this might have been the most handsome marsh harrier I had ever seen: strikingly pale, the silver dominating. I had seen him before, I think, this pale old male, for I was pretty sure he was in business around these parts the previous year. Perhaps he was already well on his way to becoming an ancestor; certainly I was inclined to think so. Did he know that three females had turned up here just the other day? And that two of them were certainly still around?

I had been fretting a little about his return – he, or some other male. Where had he been all winter? Had he travelled south, in the time-honoured manner of his kind? Or had he merely been roaming, perhaps part of that winter roost at Hickling? All I could do was guess, and welcome him back.

> Morning chores. A jackdaw tries to stop a marsh harrier doing his.

There is a barn-owl box in the bottom meadow: a nice little triangular house kindly put up by the Hawk and Owl Trust. Once again it was going to be occupied. I could see the two birds on the ledge outside, looking rather smug and proprietorial. They were jackdaws. Good luck to them. Not as sexy as barn owls, it has to be said, but birds of great intelligence and improvisational skill. I had to salute them and wish them luck.

Then – no more than a flash and a flutter in my peripheral vision – and my head turned at once, responding to things half-glimpsed and not quite seen in the usual fashion. And damn it, it was a butterfly: gallant, reckless almost absurd, but heroically leading the charge towards spring. It was, so far as I could tell from an experience that lasted less than half a second, a peacock – butterflies that hibernate in adult form and aim to make a bold early start of the real business of spring.

The real business, eh? Just like that male marsh harrier. Do they go? Know what I mean?

8

Everyone Suddenly Burst out Singing

There's always a day – one day – when it seems that the year has committed itself once and for all. The ponderous vehicle of time is now facing towards spring, the multi-point turn has been completed and—

Everyone suddenly burst out singing.

A fragment of verse that came unbidden to my mind on this morning of commitment. The world was singing.

Everyone was singing. Suddenly. Where yesterday there had been soloists, now there was a chorus. It was short of complexity, as you'd expect at this time of year, but it was marked by a sudden raw enthusiasm. There was no longer a pioneer great tit singing teacher-teacher; there were now three or four or more singing competitively, provoking each other to louder and better music. With them the dunnocks were singing out: not a great song, but sung with immense passion, a jumble of notes hurled at the turning year as if the frenzied reiterations would force the year to turn still faster. Behind them the robins are still hard at it: they've been singing all winter but now their songs are not just about territories for feeding, they're also about territories for . . . well, know what I mean?

It's when the fourth species joined in that I realised we had reached this crucial point in the year: when goldfinches were jangling and fizzing and buzzing. The year had reached critical mass. How strange it must be: most of these birds had not sung in this fine way for nine or ten months, and the young birds never, but now they were all singing as if they'd always been singing, as if there had never been a pause, as if there had been no winter, no times of privation, no period when safety and happiness lay in the flock rather than in the grand aspirations of territory and becoming half of a pair. They had not just changed habits: they had become completely different creatures.

Everyone suddenly burst out singing.

Siegfried Sassoon, of course, and though the poem is usually interpreted as a response to soldiers singing in the trenches – perhaps Welsh soldiers singing hymns in harmony – its two stanzas, its ten lines are full of birds.

... and horror
Drifted away ...

There had been a couple of days with snow on the ground and cloud ceiling a couple of feet above my head: a Chicken Licken sky. Not horror, certainly not horror as Sassoon knew horror. But as the birds – as everyone – sang, I was lighter in my heart. All at once I remembered my father explaining that some fashionable friend had her hair cut by Siegfried Sassoon and, realising his error, he started to laugh and did so as if he would never stop: he does a very nice cut and blow-dry but he doesn't half go on about the First World War. The story still comes up every now and then: and always with the same laughter, as if the joke was new-made at each remembering.

Here's what I would do for everyone who ever suffered from a moment's depression. That's a company that includes me: though, thank God, it's never got a life-throttling grip on me. I know what it is to have the blues: but I also know that the blues, when they come, will pass. And that makes all the difference. And here's my suggestion: learn birdsong. That's not a glib notion, any more than it's a cure. But I believe, profoundly, that a smattering of birdsong will help anyone who has been at home when the Dementors come to visit.

There was a time when everyone knew a bit of birdsong, as well as the words to a few hymns: it was in the nature of the times. Robert Browning has that throwaway line in 'Home-Thoughts, From Abroad' about the wise thrush who sings each song twice over,

Lest you should think he never could recapture
The first fine careless rapture!

As a neat summary of the song thrush's love for repetition, it could hardly be beaten: but it's interesting to note that the words assume a certain familiarity with the song of the song thrush. He writes as if his readers are probably aware that song thrushes repeat themselves. And yes, I could now hear a song thrush throwing his voice into the chorus: careless rapture endlessly recaptured.

It's not enough to know that there is singing going on and that it's a pretty noise and that it's a bird doing the singing. You need to put a name to the singer. Once you can identify the species doing the singing – and it's not a hard trick to master – you move to a better, happier world. Have you ever been to a forbidding party where you know no one and no one has the slightest interest in you? And have you ever arrived at a gathering full of friends and family and all your best beloveds, every one of them competing to welcome you with bigger and better hugs? Learning birdsong is like exchanging the first kind of party for the second: a gathering where you feel at home and full of love and good cheer. Learning birdsong is not just about loving nature, it's about living with nature on terms of intimacy. Being at home. The opposite of being alienated.

I had been through a few days in which I felt increasingly ancient. I was limping with increasing heaviness: something amiss with my right knee. I was thoroughly out of sorts with myself and with the world. This was not horror, neither was it despair – but when I reached the day when everyone suddenly burst out singing . . . well, it was all rather noticeably better.

Spring on the marsh, spring in my soul.

O, but Everyone
Was a bird; and the song was wordless; the singing
will never be done.

All that, and so good at cutting hair as well! Laughing again at my father's slip, laughing at the memory of his laughter as he sobbed into his pinot grigio, I got on with the day. Should I sing myself, do you think?

At my desk. If the stoat would kindly sit still, the rest of us might get some work done.

I saw a stoat, though only for a few seconds. It's said that a ninja adept can walk through a crowd without being seen. In *A Taste for Death*, one of the great Modesty Blaise thrillers, Willie Garvin, Modesty's sidekick, brings two killers into the range of his knife by means of his immobility. 'There was no expectancy in his waiting. His mind was empty, and as still as his body. Apart from the unblinking focus of his eyes, he did not exist. He was the tree, the ground, the air around him.' He stands in plain sight, and the first killer's eyes pass across him twice before he realises they have walked into danger.

Like a bird of prey, a stoat's eyes are set forward on the head: as with birds of prey, that intense arc of stereoscopic vision is the key to the way he makes his living. A rabbit's eyes are set on the side of the head: he can see all around him without moving his head. So rabbits are better at detecting movement than colour and form. A stoat travels jerkily: move and freeze, move and freeze. In that moment of frozen invisibility, who knows what creature might betray itself?

The stoat crossed the grass in front of me, freezing twice as it did so, and then vanished into cover. A stoat's power is vanishing.

And then astonishingly the stoat was on the far side of the dyke. I saw it briefly in the cleared patch of short vegetation: how splendid that this small piece of management should be rewarded. The dyke is six feet wide, Lord knows how impossibly deep, and there is no bridge. Yet the stoat, as dry and as sleek as when I had seen him before, had somehow transported itself from one side of the dyke to the other.

Some time later – it's good to be courteous to stoats and allow them time to get on with their own business – I went and checked along the dyke. I could see no place where the reeds or the brambles or the sallows leant across the dyke or met in the middle. I have seen stoats make some pretty hefty jumps, but surely the dyke would require a leap of Bob-Beamon-like athleticism. I wondered fancifully if the stoat was capable of willing himself across, of dissolving his body into constituent atoms and reassembling himself on the far side. I have sometimes had the same feeling of incredulity when watching sport: marvelling at a piece of action that didn't seem physically possible: Lionel Messi with the ball at his feet, running full pelt at an opposing player and then, without breaking stride or altering his direction of travel, continuing on the far side of the player, the ball as ever his companion. The skill is too quick for easy observation: like Modesty Blaise drawing her gun.

One of the pleasures of writing about sport, as I have done for many years, is that the action takes place in front of you, in circumstances designed to allow you a good view of everything you need to look at. But in wildlife there is seldom such luxury. This startling moment on the cleared ground was a rare example of nature as a kind of sporting spectacle. Mostly, though there is the same sense of glorious action, you never see all of it, you never know all the circumstances, and quite often you have no idea who won and who lost: only that the action had you enthralled.

I have often been asked what sport and wildlife have in common, since I make my living by writing about both. I have never come up with an adequate answer. The nearest I have got is to answer: nothing at all, obviously – that's the whole bloody point. But I think I might have caught onto something when I began to keep a notebook about magic, wondering why on earth we require the concept of magic in our busy, rational, 21st-century lives.

Time and again I have heard words like magic and magical used to refer to great moments in sport and in wildlife, to great individual performers in sports, to extraordinary, charismatic species. Sharing a dawn with a herd of elephants or swimming with dolphins is generally described as a magical experience. Virat Kohli, after a great innings, might be described as 'a magician'. Jose Mourinho's skill in motivating players is routinely referred to – in 'serious' newspapers – as magic.

Magic is not what we can't explain but what we don't want to explain. Both sport and wildlife are reckless – profligate – in supplying this kind of magic.

And it's not something any of us want to live without.

Sunlit morning walk. Winter raises the white flag: a brambling's bum.

Sometimes people who suffer from depression – the real thing, the fully Monty – end up at Clinks, sometimes for a while, sometimes as part of a routine that keeps the blues at bay. Hard physical work is a good thing, but hard physical work in the open air in all weathers on behalf of animals has always seemed to me a deep privilege. Even on the worst of days – bad weather or the blues – the horsey chores tend to put me in a good temper.

There is an extraordinarily pleasant atmosphere at Clinks. It's something I've noticed a lot when doing stuff with Eddie. Being in such places is one of the privileges of being Eddie's father. At Clinks, at other places Eddie goes to – his swimming group, his drama group, his riding group, social events with his friends – there is an atmosphere of easy tolerance: a sort of man-of-the-world acceptance that people will be different and that they will do the things they do. (I should note here that Cindy does about 100 times as much as I do in this sphere.)

There is a natural tendency for the fortunate to feel ill at ease among the less fortunate: to feel alarmed, imposed on, unsure of themselves when forced to deal with a person with Down's syndrome or autism. That is surely a trifle self-indulgent: after all, it's the people with the problems who have the problems.

But when you are in places where such difficulties are taken for granted, you find this astonishingly pleasant vibe. It's like being at a gathering with a lot of famous people: there is a tendency for those who are not famous to feel unsure of themselves, uncertain how to treat this apparently different race of people. The answer is the same in both cases: treat 'em like human beings. It's easy: it's just that it takes a little getting used to.

After Eddie was born I no longer had the luxury of leaving disadvantaged people in the hands of saints. I had to get stuck in myself. And blow me down, behind the conditions and the difficulties and the disabilities you find human beings. I have helped out a few times at Eddie's Riding for the Disabled group – RDA is a great thing, for an astonishingly small sum it's possible for people with considerable problems to get on a horse and for once look down on the rest of the world. I have helped those with uncontrollable twitches, a silent girl locked up in herself but finding real release in her communion with her horse, an autistic lad who required constant conversation, another who required two or three repetitions of every instruction. And it was easy, not just because I am used to being around Eddie, but because the essential vibe of this sort of gathering makes light of something that would be a big deal in other places. To spend half an hour with a troubled child: well, it would have been an ordeal 20 years ago, but I am a different person these days. There has been no option.

These places are valuable for those who have the problems: they are also pretty good for the people who do the helping. Perhaps the most shocking thing I have learned from being Eddie's father is the unexpected number of good people there are in the world: people with real spiritual generosity. Eddie sometimes attends a kayaking club, one not specifically for people with disabilities. The instructors were hugely generous with their energy and their time, and seemed to understand Eddie's needs more or less at once. The process seemed to give them as much satisfaction as it gave Eddie. He made their day better by being who he is.

There's a link with this phenomenon and with the wild

world: you can't fake the real thing in either. I have travelled widely to write about wildlife and conservation, and again and again I have found people to share the wonders with. Out there, in rainforest and mountaintop and savannahs and desert and inner-city park, I have met people and felt that instant good vibe. The shared and unfakeable love of nature, being in it and helping it to carry on. There is the same unmistakable good vibe at places like Clinks.

So here is a gloss on my remedy for depression: learn birdsong, yes, but if possible, learn birdsong with Eddie as a companion. Don't take a field guide, take Eddie.

'What's that?'

We were on the marsh, it was evening, well wrapped-up: cheese sandwiches and apple juice and beer.

'You know that one. But it's a while since you last heard it.'

Again: a bark with a slight break in it: a hard, strong monosyllable.

'I'll give you a clue: they nest in that wood over there, all together—'

'Heron!'

Much praise. And much joy: it seems that the pigeon-shooters hadn't put the herons off the place altogether. We could see two of them on the marsh, ghostly grey shapes standing still in the way that herons do: stillness is a heron's power. They have two speeds when it comes to fishing: none whatsoever and warp. Stillness works for them as it does for a stoat: when they're motionless they are invisible to a fish: a grey cloud in a grey sky. They will, very slowly indeed, increase their angle of lean, till it seems they must fall beak-first into the dyke. And then the stab, too fast to see, like Modesty's gun: what you generally see is the recovery stroke, often enough with a fish athwart in its beak. Herons don't use the great beak to stab with: it's a not a spear, it's a grab, and the length of the bill is the margin for error in the strike. It's as if the gun appeared by magic in Modesty's hand; it's as if the fish leapt sideways into the heron's bill.

And so I rejoiced for the herons, though as always in these cases, it's possible the fish had another view.

When you live as a family – when you have that immense privilege – everything that happens is about the family. Writing, artwork, music, school, Clinks Care Farm – everything. That includes the life of the marsh and its management. A few years earlier, the plans we had for the marsh had to be put on hold: Cindy's sister Cherie was going through a long and eventually fatal illness, and Cindy was constantly commuting between home and Cherie's flat or hospital in London. Joseph went through a health scare. Eddie went through great troubles at school: there were times when he went into lockdown and could neither speak nor move, and the teachers – some of them – thought he was just acting up. They acted as if he was deliberately misbehaving: write out 100 times: I must not have Down's syndrome! And me, well, the freelance life is never straightforward and some days were better than others.

All of which meant that, although the marsh gave us all, in our different ways, great comfort and great meaning, it didn't always get the great management it would have had, had it been run by a major conservation organisation. The place was inevitably affected by the ups and downs, the mood swings, the fortunes and misfortunes of the family that owned it. There were times when it had no choice but to get on with the task of being itself: something it was able to manage in a very satisfactory manner.

That's how it went, that's how it often still goes: we want to clear scrub to create new habitat, but I have a deadline to meet and then I must organise supper for Eddie because Cindy needs to be with her sister. And so family life keeps on keepin' on in its own wild way, while the marsh carries on doing the same thing.

> Song thrush at his morning workout. How many reps today?

When we moved in we found ourselves rather unexpectedly the owners of half a dozen enormous koi carp. They lurked in the dykes that border the garden, their egress blocked by judiciously placed chicken wire. We made a few attempts to catch them because such fine ornamental fish would give great pleasure to those who love them. They are also worth decent money: some top specimens go for more than £500. Even if ours weren't in that class, they would be worth a few bob, so Cindy got nets and buckets and tried to persuade them to leave the dyke. But they were canny things, or perhaps not so canny: a friend came with a great net to whisk them away, but they at once sank down to the bottom and became invisible.

This was a short-term success for the fish, but a long-term error. The people we bought the house from had a taste for large things: if they liked something, they liked a lot of it, and as big as possible. So as well as the giant koi, they also had three colossal dogs of immense vigour and volume.

Their presence was enough to keep the wildlife of the marsh restricted to the marsh, but after we moved in the boundary between garden and marsh grew less defined – and a lot more quiet and a lot less bouncy.

It was, I think, during our first spring at the place, that we looked for the koi and we looked in vain. We checked several times after that, until we were sure of it: they weren't playing coy, they had gone. It's possible they had staged a mass break-out and gone feral, but I doubted it. And they were too big even for the most ambitious heron. So it was almost certainly the otters that got them: no longer with the dogs to worry about – their oppressive presence, the sound of their barking, the lingering scent of their existence – the otter could come into the garden without fear. Several times I found droppings near the house: otters will often travel across land between waterways. And I suspect that these great, fat, charming, slow-moving fish had formed a banquet for the creatures that are really the seldom-seen soul of the marsh.

> Morning ride. In the sunlit stubble a hare thinks about going mad.

Those who live urban lives seldom if ever experience the joys of the country store. Which is a great loss. They're huge and draughty and cold and full of immense hammers, wheelbarrows of Brobdingnagian dimensions, sacks of feed for every known domestic mammal or bird, and all kinds of clothes suitable for the most reckless forms of reckless dressing-up. In our favourite, New Atlantic Country Superstore, there are also tanks of fish to marvel at – and should we wish to please the otters, there are plenty of koi for sale. When I enter such a place on legitimate business my boyhood in South London seems not history but deranged fantasy.

Eddie and I were there to get new boots. I favour construction boots for work at home: you can pull them on and off without bother, and they put up with all kinds of abuse. They also have steel toecaps, and if you have ever been trodden on by a well-shod horse, you appreciate the mental comfort the steel brings. My current pair had gone beyond the picturesque distressed look and were now shipping water in quantity; walking around the marsh in such footwear had become a suboptimal experience.

Eddie needed a pair of boots to wear at Clinks as well as around the stables and the marsh. We found a pair that was waterproof and with Velcro straps, which meant he could take them on and off himself: good for now and also, as you never stop wondering about, it's a good step for the future, and greater independence.

My own new boots were dark brown and lined with fuzzy fake fur: they slid on easily and, once on, felt good and solid. Eddie wore his boots out of the shop.

'Do the horses tonight,' he said. 'Snack on the marsh.'

'Getting late,' I said. 'Snack on the marsh tomorrow.'

As I finish the evening chores the heronry barks a good goodnight.

Those koi we had seen in their big tank at the country store, looking up at the light with such mournful expressions: should I regret the part they had played in the otter's banquet? The other day I had seen a sparrowhawk – neat, small and dapper so certainly a male – working the edge of the dykes. Sparrowhawks are not much loved by the tender-hearted.

I remembered a piteous description of a sparrowhawk's successful raid on a generously supplied garden full of bird feeders. I had written a generally approving piece about sparrowhawks: this letter was supposed to be a refutation. The sparrowhawk had caught a starling and devoured it in full view, and the cries of the starling, who died comparatively late in the course of this banquet, had pierced the writer's heart. Such cruelty should not be permitted.

And yet the sight of that lithe male sparrowhawk, flying so low to the banks of the dyke that it seemed impossible that his wings wouldn't strike the ground with every downstroke, seemed to me a wholly admirable thing.

Later that same day I had seen one of the female marsh harriers making a low pass over the grazing marsh that lies beyond our own chunk of land, and 20 curlews had leapt into the air in response, calling boisterously. I could see the long decurved beaks, the sharp white vee on their backs. Should I cheer for the curlews who got away? Should I condemn the marsh harrier for disturbing them? They are both, in their different ways, magnificent birds: do we have to make a choice about what kind of magnificence we like best?

Predation is an emotional business. I have many times been with people who have wept bitterly and inconsolably when witnessing a lion kill. They all tucked into their meat when they returned to camp.

On the marsh, as on the savannah, death is everywhere. The otters, the harriers, the sparrowhawks depend on death for their lives. So do the blackbirds for that matter, so do the robins and blue tits and great tits and wrens.

Nothing to get blasé about. Nor in the main to weep for.

Morning chores with a train to catch. Hurry up, says the song thrush. Hurry up, hurry up!

Sometimes on evenings in late winter and early spring, blackbirds will chink. There's a curious ambiguity about chinking. When they settle down in a spot to roost for the night, you'd have thought that silence was essential: after all, you don't want to tell every predator of the night that you'll be helpless for the next 14 hours. But blackbirds chink, and as they hear the chinking of others they chink back. It's about joining in. It's about being together – and yet not being together. It celebrates differences as well as unity, for there is something competitive in it. There is a hint of territorial behaviour, but also a sense of group solidarity. It's a contradiction.

You can see the same kind of restlessness when any species of bird settles into a roost. They are competing for the best places – safe in the middle rather than isolated on the edge – and that's a slightly grim business because they are competing as to who is least likely to be eaten or to die of cold in the course of the night. But at the same time they need each other for their own protection: the more birds there are, the safer each individual becomes. The birds are competing and co-operating at the same time.

As the days begin to get longer, so the chinking chorus becomes more pronounced. Every day as the light fades, the chinking strikes up: an insistent monosyllabic rhythm. That's another advantage you get from looking after livestock: as you go about your evening chores, so you are perpetually tuned into the routines of the world going to rest.

I remembered a pub in St Albans. It was a Thursday: apparently it was the custom for Irish musicians to turn up at this place with fiddles and guitars and a bodhrán – Irish drum – or two. So I drank Guinness and sang along with 'Whiskey in the Jar' and 'Black Ribbon Band' and very jolly it was too. Eventually, the songs were sung: closing time was nigh. Bairbre, who had taken us to this place, was speaking to some of the musicians. The show was over, they were putting instruments back into cases and restoring ravaged throats with a final sluicing of the good black. All around was the lively conversation of 11 o'clock. Then suddenly – and oh, my brothers, it was like some great bird had flown into the bar – a single voice rose in song, high and pure and marvellously sweet. No instrument offered support: just a single human voice, singing in Gaelic. It was Bairbre – of course it was Bairbre – and all around, the voices faded into the silence and the song rose and fell and died away.

It was like that as I put the horses away that evening. Late winter, early spring: and from the clatter of the chinking chorus a lone voice rose up and sang: not a chinking but a full glorious song, as much like the chinking as Bairbre's song was to normal conversation as she gave us 'The Green Fields of Gweedore'. The first blackbird song of the year: you might think it was an appropriate moment for cheering and shouting, but I did nothing of the kind. For a start, it's unwise to make sudden noises when you're leading a horse, and for a second, some pleasures go too deep for shouting.

I have been to sporting events that have been so marvellous I had no desire to cheer at their conclusion. I wished only to sit and marvel at the moment of marvellousness. Again, it's like religious awe, even though it is nothing of the kind: a moment to savour things deeper and more marvellous than anything I could ever be myself.

Bairbre is now Eddie's godmother – his God-mammy, we jest. Oh, and the line about 'oh my brothers' is from *A Clockwork Orange*: the moment near the beginning when a voice rises above the din of the Korova Milk Bar to sing a few bars from Beethoven's Ninth.

> *Alle Menschen werden Brüder*
> *Wo dein sanfter Flügel weilt.*

All men become brothers under your tender wing.

The blackbird's sweet song briefly filled the night. Let the hostilities and rivalries of spring begin. Blackbirds can be brothers again come the autumn.

> How young am I this sunlit morning? I can still hear the goldcrest.

Never make promises to Eddie. He remembers them. So we were out there again: late afternoon, the snack – slices of the excellent banana bread we had made over the weekend – as promised.

As the light faded, a barn owl headed straight for us, that white heart-shaped face staring us down, the owl curving off as he realised he had company on the marsh that evening. Silent, always silent. Then a wise thrush, singing each song twice over: still early enough in the year to be a thrilling novelty.

And then for the first time that year, the sudden shout of

a Cetti's warbler. For me there was an ocean of relief in that buoyant cry, and something almost comic in its assertiveness. One of the things I love about Cettis is that you never see them. Hardly a glimpse. They are massively talented at keeping out of sight, and at the same time massively talented at making sure you can't miss them – when they don't want to be missed. Cettis! Sometimes it sounds like onomatopoeia. I have a favoured mnemonic for the call:

Me? Cettis? If-you-don't-like-it-fuck-off!

That gets the rhythm across, along with the essential message, so far as other male Cettis are concerned.

It has always seemed to me to be the voice of the marsh: the identity of the place contained in those cheery life-demanding notes.

'Do you remember that one, Eddie?'

He wanted to, but the name had gone. 'Ch . . .' I offered. 'Ch . . .'

'Cettis!'

'Brilliant boy!'

Spring was here all right.

9

Not Dying

Doris has had her magical way with the marsh. Two willows . . . whomped.

My old friend Darrell was talking about his knee replacement, an operation required after years of tennis. I countered with my right shoulder damaged by years of keeping wicket and goal. 'You realise,' said Darrell, 'that we'll be having this conversation for the next 30 years.'

'Only if we're lucky.'

Ageing is a form of living, not of dying. Limping has its drawbacks, but it's a form of Not Dying. And that always has something to commend it. But my right knee was getting worse. This would occasionally cast me into fits of gloom: had the best of it, now for the endgame, bits dropping off, never be the same again.

There's life in such grumbling. I would limp round the marsh, increasingly vexed, sit laboriously, swear a bit. I was feeling Vulnerable, or perhaps Near Threatened – two of the categories for endangered species used by the International Union for the Conservation of Nature, as it happens. Still not Critically Endangered, not yet. And the place itself, threatened yes, by the forces that surround it, but it still had marsh harriers flying over it.

Winter was fading fast, as spring advanced and retreated, advanced and retreated. A day of fierce chill made the blackbird's glorious solo seem like a false memory: a deluded fancy of some poor and inaccurate observer. But these cold days were not about dying. They were vigorous days. Full of life.

I was reminded of David Hockney's quartet, 'Three Trees Near Thixendale': the same view in four seasons. In the winter one the trees are bare all right, but the picture is beaming and gleaming with colour: blue, orange, green. Winter is full of life. It lacks the making-babies urgency of spring, but it's thrillingly vivid in its simple determination to survive.

Woody Allen famously said that he didn't want to achieve immortality through his works, he wanted to achieve immortality by Not Dying. 'I don't want to live on in the hearts of men, I want to live on in my apartment.'

Winter on the marsh is a bit like that. Soon enough the place would be bursting and booming with new life, insects everywhere, butterflies, reptiles, bats. But right now it was still mostly engrossed in the task of Not Dying: and that negative makes a glorious positive. Sure, many individuals had already died in the cold, and more would die yet, but that's expected. The natural mechanism of life takes death in its stride: that's how it works; that's how it's always worked.

Spring always seems more distant as it gets closer: its setbacks are sometimes almost unbearable. But if you look at wildlife a lot, you dwell on the fact that every landscape continues to live: that it's operating its strategy for not dying. The fieldfares that came here for the winter were not dying of cold in Scandinavia. The bats that were hibernating out of sight were not dying of starvation on the marsh. The torpid reptiles, the insects in diapause: they had drastically slowed down the processes of life in order to get on with the job of Not Dying. The kestrels hovering and hunting over the marsh – the part we bought from Barry is especially important to them – were finding enough food to maintain their high-energy lives.

Winter is a time of death. Like all the other seasons. And like all the other seasons, it's also a time of life.

Far away, on the grazing marsh beyond our boundary, I could see two swans. A storm of white wings. They copulate on water, but right now they were displaying: assuming heraldic postures and thundering their wings at each other. For them spring was here: they'd got through the winter, they'd already done the Not Dying bit. The marsh, the landscape and the swans themselves were moving on.

Every day it changes, sometimes clearly progressing,

sometimes appearing to fall back. I, greatly cheered, limped back to the house, grumbling and swearing.

> Morning ride and the skies are full of song.
> Such larks!

The first butterfly of the year is always a good moment. More than any other creature, a butterfly is light and warmth. The life force of winter is all very well, but you forget about such notions the instant the first butterfly flutters by. The other day I correctly identified a sparrowhawk without seeing it – at least, not in any demanding two-eyed sense of the term – or hearing it. It was more like awareness: something in the patterns of life beyond and slightly behind me, and my mind instantly made the connection with sparrowhawk. They're sudden birds, more often seen from the tail of your eye than with the full pomp of stereoscopic vision: your brain stores that pattern and when you encounter it again, it interprets scanty information in a meaningful way and ... well, *bam!* You've seen a sparrowhawk.

Or a peacock. A peacock butterfly, for that's surely what it was. Could I really count that earlier occasion, the briefly seen possible or probable peacock? There was something about the size and the boldness of its flight, and perhaps its darkness as well. And of course, a little knowledge. I knew that peacocks overwinter as adults (rather than eggs or caterpillars or pupae) and so when the triggers of warmth and light come, it can get on with business right away.

It took me a while to start noticing butterflies. Amazing that there can be so many beautiful things flaunting themselves in front of you, and you just say 'oh-ah' and carry on. I remember looking at a peacock supping from a buddleia and wondering if this wasn't some fabulous rare creature, something I was specially privileged to see. Well, I was half right, I suppose. It was a classic example of blindness to the commonplace: you see something so often you never actually see it at all.

They do have an exotic vibe to them: four great eyes staring back at you. We have all seen incredible, damn-near-impossible examples of mimicry in the rainforest, most of them shown to us by the great David Attenborough: well, here's one just as good, and it's in your back garden and along the railway line. A peacock, when seen upside down, turns itself into the face of a small but very cross-looking owl: little points in the trailing edge of the wings look like ear-tufts, the body of the insect looks like a beak, and the great staring eyes do the rest.

Underneath, the wings are black: the colour of the dark crevices a peacock will lie up in. But when it's discovered by a mouse or a blue tit, the peacock will open its wings and present the predator with the mad face of an owl, at the same time rubbing its wings together to produce a hissing sound that's just like an irritated owl. It's as if your Sunday lunch shape-shifted into a psychopath with fixed bayonet.

There's always good news and bad news with every early sign of spring, especially one as fragile as a butterfly. Was this a reckless pioneer brilliantly stealing a march on the rest of his kind? Intra-species competition and all that? Or was it a doomed individual, finding a tryst with no one and nothing but cold death? One more of those stories with the last page missing: an eternal part of all involvement in wildlife. So I feared the worst and hoped for the best.

Perhaps the butterfly felt the same thing.

The goldfinch's song sounds like a fruit machine paying out. Jackpot! Spring!

Just across the dyke that forms the boundary on the far edge of our bit of marsh, there's a grazing meadow with a short sward. It's an exposed spot; a person unused to East Anglia may even call it bleak. And it had become a great favourite of curlews and lapwings, who often feed and roost there, as the marsh harriers know. It was a fine thing to see a flight of 40 curlews, with their long and stylish decurved beaks. They take their scientific name from this: *Numenius arquata*. The first part means new moon, for the crescent of that bill, and the second is archer, for the bill's longbow shape: so here were the archers of the new moon, dropping down in the distant field all together in a flock, rather in the manner of water going down the plughole.

It's the call that gets to you: they say their own name with a strong stress on the second syllable. It's a call that fills the sky – and there's a lot of sky to fill round our place. The sound is wild and far-carrying, and in human terms it seems at the same time wonderfully sad and gloriously defiant. They are a perfect example of the liveliness of winter: that inspirational call encourages us all to keep buggering on. They spend the winter mostly on coastal wetlands, where they can forage, and in the warm weather they go to the uplands to breed: the lowest of all birds or one of the highest.

They are Britain's most sharply declining bird species, and the competition for that accolade is pretty intense. They went down 46 per cent between 1994 and 2010. There's a suite of possible reasons for this, still being investigated: these include changes in agriculture and forestry on the breeding grounds. So when you see curlews in decent numbers it's another classic good-news-bad-news moment: good to see them, bad to be reminded of their decline.

And for me, there's also a personal thing going on.

The curlew group – 13 species, the Numeniini, or new-mooners – is rapidly declining across the world. One reason for this is the destruction and degradation of coastal wetlands. They are one of the most threatened groups, though again, this is a hotly contested title. The Eskimo curlew was last seen in 1963 and is more or less certainly extinct. What tends to happen, in these increasingly familiar events, is that populations cease to be viable and the species no longer functions as a continuing part of our planet, but one or two individuals hang on – sometimes referred to by scientists as the Living Dead.

The slender-billed curlew was last seen in 1995. A few years earlier, I made a trip to Morocco with my old friend Martin Davies, then of the RSPB, known in the birding community as one of the co-founders of the annual British Birdwatching Fair at Rutland Water. It was an excellent trip: good birds and good craic. One morning we got up at dawn to look for curlews. We found a dozen or so: the Eurasian curlews that I see on the marsh back at home, and among them an oddity: quite different in build, in what birders call 'jizz', the vibe or feel of a bird, something that comes from its shape and especially its movements. This was a slender-billed curlew. Now almost certainly extinct. That bird Martin and I saw was a member of the Living Dead.

So I have seen an extinct bird.

That's not something to take pride in. I wish there were hundreds of slender-bills out there, thousands. I wish they were on every birder's list. I wish that slender-billed curlews were so common that no one even bothered to boast about them – might as well boast about seeing a sparrow. As it is, I have the melancholy distinction of having seen a bird that we've managed to wipe out.

And that's a haunting thought. I can still see that neat, elegant bird picking its way fastidiously through the wild flowering cresses of this soggy piece of land, feeding contentedly as if the world held nothing that could ever be a worry. So when the curlews drop in and pay us a visit at home on the marsh, I am reminded of that lone bird, one of the last of his kind, walking among the pastel colours of the cresses as if living an idyll: a lovely gentle world in which nothing could possibly go wrong.

> As I complete the morning chores the dunnock continues his. UDS: Unilateral Declaration of Spring.

Father Francesco Cetti was an Italian Jesuit. He was also a naturalist and a mathematician. He was born in 1726 and wrote *Storia Naturale di Sardegna,* The Natural History of Sardinia. I got these facts from *Whose Bird?,* which lists all people who have had birds named after them, and supplies a brief spot of biography. You can read about Bewick, of Bewick's swan, who was the great print-maker and ornithologist; and about Montagu, of Montagu's harrier, the 18th-century British naturalist: both British birds, and both birds I would love to see on the marsh.

British birds are a little light on birds named for humans: we have been aware of them too long and their folk-names have become part of our language. But if you travel, you find many more birds named for half- or completely forgotten men, and a few women too: there's a Mrs Moreau's warbler, an African species that was described by Mr Moreau. It was rediscovered by my old friend Bob. He communicated his findings to Mrs Moreau herself, who was by then a widow: a touching tale.

I have seen Pel's fishing owl and Heuglin's robin and Arnot's chat and Bonelli's eagle and Böhm's bee-eater. There's a Bonaparte's gull, named not for the emperor but his nephew, who was an ornithologist. Thrillingly, there's Barnes's wheatear, which is a subspecies of Finsch's wheatear; this Barnes was a professional soldier who rose from the rank of private to become a commissioned officer; he was also an ornithologist. There's also Barnes's cat snake in Sri Lanka, and there's a species a chiton – a kind of mollusc – called *Radsia barnesii*.

My favourite of these name-bearing creatures is the lovely Narina trogon, which the French naturalist François Levaillant named for his Khoikhoi mistress. The word means 'flower' in the local language. It was not her given name, but it was what he called her: so we have a lovely bird that bears the love-name of a beloved lover.

There are 18 species named for David Attenborough, which is all very right and proper. I hate it when the names of species are rationalised and reorganised for no very good reason save busybodying. These days I'm supposed to refer to Heuglin's robin as a white-browed robin-chat: well, when I take clients into the bush in Zambia, they have to accept a few archaic names. Heuglin was a 19th-century German explorer and ornithologist, he has eight species of bird named for him, and I think it's right to keep his memory going with the names of his birds. These names matter: they are part of the long and complex history of human interaction with birds.

The scientific name of Cetti's warbler is *Cettia cetti*: the genus *Cettia* is in the family of Cettidae, which contains a couple of dozen species. So there's glory for you.

So I'm always ready to raise a glass to Padre Cetti. Imagine having your life commemorated by something as wonderful as a living bird. In its way it's a form of Not Dying.

> Morning chores. It seemed an unpromising sort of day to me but what do I know? The blackbird sang out anyway.

The miracle of the shape-shifting peacock butterfly – the butterfly that becomes an owl – is a classic example of the marvellous nature of everyday life. How many miracles can you find in a single back garden? As many as there are living things, I suppose. When our eyes pass over something as if it were beneath our notice, we demonstrate only our ignorance. The plain dunnock, belting his plain old song out from the bottom of a bush, is no one's idea of a thrilling bird: especially not in a place where marsh harriers roam and Cetti's warblers shout out their songs.

But the passionate and violent nature of dunnock life has been made clear in recent studies, and they have almost overtaken the robin as everybody's favourite example of the barbarity of non-human life. Robins have a revisionist reputation for uncompromising violence, but with the dunnock, it's about uncompromising sex. Here is a brownish bird with greyish head, and a rather monotonous song: and yet, while all this singing is going on, there are multiple infidelities and the violent routine of cloacal pecking – a direct assault on the genitals.

All birds, amphibians and reptiles have a cloaca: a single bodily opening for the purpose of urination, defecation and reproduction. A few mammals have the same system: platypus, echidna, tenrecs (a diverse group of mammals found in Madagascar and some parts of the African mainland), golden moles and marsupial moles. Dunnocks have an intense taste for extra-pair copulation – what we humans call infidelity or cheating – so copulation tends to be preceded by cloacal pecking: the male deftly removing sperm deposited by previous partners.

The moral of that, I suppose, is that no species is boring, and that every common thing is exotic. We humans have always loved contradictions. We rejoice at the security and comfort of the familiar; we also thrill to the exotic and unfamiliar. Marsh harriers cross the marsh more or less every day: I have watched them do so in sunshine and snow and rain, at midday and at dusk. It is a deep and rich privilege to live alongside them.

But a peregrine passing through is a special excitement – and for the exact opposite reason. They are occasional birds here; they acquire glamour by means of their absence. They also have the swankiness of being the fastest of all flying birds: in a vertical plunge after prey – flying birds – they have been timed at 200 mph and there are still more fabulous claims. In a stoop they not only use gravity: they power vigorously into the descent, gaining speed as they hurtle down, and when they strike, the targeted bird is likely to die from the sheer shock of the impact.

There's also a shape-shifting thing going on with peregrines, though unlike the peacock butterfly, it's unintended and exists entirely in the human brain. Many times – perhaps every day – I see a wood pigeon and fancy for a second that it's a peregrine: they can be confused so long as you see the bird badly enough. And often enough I have seen a kestrel and wondered for a moment if it wasn't a peregrine: both are falcons, one the hoverer and the other, the bolt of doom.

Sometimes it works the other way, though rarely. I once saw a pigeon take wing from the Tate Modern – I was on the Millennium Bridge – and as I watched, it turned into a peregrine. And the same sort of thing happened out on the marsh: I raised the binoculars to enjoy a kestrel and in an instant, it had revealed itself as the burly speedster with the cad's moustache.

I wonder: would Gerard Manley Hopkins have written a poem even more brilliant than 'The Windhover' if he had seen a peregrine on his morning walk, rather than a kestrel? Would the greater exoticism of a peregrine have stirred him to still greater heights of genius? Or was the comparatively homely nature of a kestrel essential to the perception of the wonders he saw in and beyond the kestrel:

> *dapple-dawn-drawn Falcon, in his riding*
> *Of the rolling level underneath him steady air*

Two things. He wrote the poem in 1877 at the height of the Victorian persecution of birds of prey, so perhaps the kestrel really was a mad exoticism so far as he was concerned, while peregrines were so well shot up as to be unobtainable. And secondly, a hovering kestrel makes a cross. Sometimes they hover without moving their wings at all: gliding at the exact speed of the wind they are facing. So perhaps it was a combination of the exoticism of the kestrel and its apparent eagerness to become a religious symbol that allowed him to write that most ecstatic poem.

I have often felt the ecstasy in the falcon's flight – when reading the poem and when seeing a falcon in action. For me too the falcon is a symbol – of falcons, of wildness, and perhaps of hope.

I went up to the workshop to tell Cindy about the peregrine and found her completing a gorgeous kestrel, its outline cut from wood. It was painted in gold-vermillion, and surrounded by the words of Hopkins in wild calligraphy.

> Riding out, we caught this morning morning's minion.

There were lions on the marsh.

Two females, resting up under a tree as lions will. They were lying on their breastbones, as if in Trafalgar Square, front paws extended, heads up, inspecting the lion-coloured vegetation of mid-March.

For a good length of time – maybe half a second – I was convinced of this. I could see their shapes, as beloved as they are feared: I could see their eyes, dark in their sandy faces, I could see their whiskers, I could feel the glorious, confident, slouching manner in which a lion faces the world.

The double-take is a standard piece of comic business. A character sees something and then makes a violent adjustment to look at it again: for what he has seen is shocking, unbelievable, impossible. There's a YouTube clip of Patrick

Stewart teaching the quadruple-take: a fine demonstration of physical comedy.

So naturally, when I saw those two lionesses on the marsh, I performed a double-take – maybe even a triple- or a quadruple-take. My eyes had sent a visual message to my brain, and the brain had interpreted it. Wrongly.

The lions were a classic error. When I had completed my double-, triple- or quadruple-take, I realised that the lions were fallen trees, part of the work of storm Doris.

Doris had struck us a couple of weeks earlier: a ferocious wind that took down half a dozen of our trees. Those that fell on the marsh were left to rot or regrow. When I show people round the marsh, there's fallen willow I always point out with immense pride as the only argument with Cindy that I have ever won. The tree fell during our earliest days here. She suggested that we get it cleared and chopped for firewood; I said we should leave it to lie there and do what it wanted. It's now a cheerily living if horizontal tree. It was out there Not Dying and was already putting out this year's new shoots. A Suffolk farm-worker once told me: you can't kill a willa. This tree had no interest in death.

But there was a dead willow on the marsh, and one I had always had a rather good feeling about: a landmark 12 feet high. I remembered Norman Stills, who converted a vast carrot field into a fen for the RSPB. During this almost magical conversion of Lakenheath Fen, he found an enormous chunk of bog oak buried in the earth: long-dead and preserved by the peat. So he had this set up on end, to provide a kind of eminence, declaring: 'One day an osprey will perch here.' And eventually one did.

I had the same ambition for the 12-foot willow stump, but

it was now a six-foot willow stump. Doris had whacked it in half. No osprey had ever perched on it, to my knowledge, but I had always enjoyed the thought that one day one might. Would this much lower perch be quite as attractive?

My eye, or rather my brain, had not yet completed the job of storing the images of the marsh in its post-Doris state. So when I glimpsed, in my peripheral vision, the sun catching a pair of stout new-fallen branches, my brain supplied not the image of the marsh, but of the savannah: of the Luangwa Valley where I had been five months earlier. It was like the experience of greeting someone you know slightly and then realising you're talking to a stranger.

I could feel my brain's embarrassment, and its quick covering-up of its faux pas, replacing this preposterous image of Norfolk lions with one of fallen willow branches. I felt a pang of tenderness for the beloved Valley, and with it I felt a strong sense of remembered fear. Me and lions, we go way back.

The wood says chiff and the wood says chaff.
The overture is over: let the concert begin.

As I have cheered for a goal, for a wicket, for a double straight somersault dismount, for a victory, so I have often cheered for wildlife. It's one of the deep human instincts: that moment of exaltation in which you can only just bring yourself to believe what's happening before your eyes. 'My heart in hiding stirred for a bird,' wrote Hopkins of his kestrel, and my heart was stirring for everything that spring brought us.

The blue whale!

David Attenborough's great soul-cry as the gorgeous

monster from the deep came to the surface right beside his tiny boat: his joy instantly becoming everybody's joy, especially Eddie's. That too was a form of cheering: of shouting hurrah at the marvellousness of the living world. I once told him – we both have an association with the excellent charity, World Land Trust – that when I met my own blue whale, I wasn't anything like so eloquent.

Fuck!

'Well, I *thought* that,' he said generously.

Eddie's love of the blue whale began with that classic Attenborough moment. We watched it again one evening – it's a regular treat – and he remembered our visit to the blue whale of South Kensington a few weeks back. So he decided to write a poem for the man who has inspired both of us. Cindy put it in the post, and a week or so later we received a letter – handwritten, that being David's generous way – thanking him. That too was worth a cheer.

There were also primroses to cheer, little yellow suns lurking in hollows and dips by the roadside. A red admiral, then another. Then, tearing another Molly-Bloom Yes from my throat, a brimstone: a butterfly in pale yellow, the colour of butter, in fact, and so giving the name to all of his kind. I saw a bat on two separate occasions, one small, the other large. That's almost as deep as my identification skills go with bats: I suspect the small one was one of the three pipistrelle species found here, and that the other was a Daubenton's, a species that loves water.

I kept seeing herons on the marsh. There was a comical sighting, just the head visible, the rest of him concealed in a dyke. Increasingly I saw herons flying towards the wood, and occasionally I saw activity going on within the wood.

There are a thousand special moments in the spring – each one, it seems, a lethal blow to the winter. But there are one or two signs that seem to mean more than the others. It's a human choice, a personal choice, depending on what kinds of wildlife you look at most. Ralph once told me that I looked at the wild world in the manner of a cat: I don't notice anything unless it moves. 'Or makes a noise,' I corrected.

I know what he means: for him early flowerings and sproutings tell him more about the turning of the year than any bird. But for me, there is a big moment that comes with the first chiffchaff: and there he was, hammering out his simple song as if calling the world to sanity. Chiffchaffs are warblers. They migrate, but only to southern Europe and northern Africa, and so they are usually the first migrant that returns to our land: that's worth a special cheer, it seems to me.

These days, with climate change, increasing numbers are overwintering in Britain, but that fact doesn't dilute the glory of that first chiffchaff, saying its name again and again.

The snag was that my knee was worse. Walking was becoming less attractive, or indeed possible for any serious distance. I was still able to ride: that didn't seem to put any stress on the knee. I was riding home on a fine morning of early spring but, despite that, I was glooming away to myself, wondering how much longer my body and my nerve would hold up. How many more times will I be able to get on a horse?

A horse can jump six feet sideways in about half a second: a fact I wish more drivers were aware of. Miakoda, though strong and confident in traffic, will every now and then perform a spook for no reason that's apparent to her rider.

We were about 200 yards from home after a nice ride – spring, chiffchaffs, primroses – when she did it. She was walking, I was holding the reins by the buckle, having deliberately passed the controls to her. (If you never trust a horse, how can you expect a horse to trust you?) I was pondering my misfortunes and riding with a slightly vacant expression, relaxed and deep in the saddle, when she jumped. Here's what I did: absolutely nothing whatsoever.

Well, I must have used the muscles required to stay in balance, but my position on the horse remained exactly the same. One second we were walking along the left-hand side of the road; the next – no traffic on our lane – we were doing the same thing on the right. I suspect my slightly vacant expression was also just the same: like Buster Keaton walking through the landscape of collapsing buildings. It was one of those powerful messages from the non-human world: for God's sake, shut up whinging! It's bloody spring!

Deeply soothed by this reprimand, I rode the rest of the way home and put the horses out into the field. Was that a flush of new growth in the grass?

> Morning ride. My horse is swift but she can't outrun the peregrine.

Eddie wanted to record the lunchtime species-count in his diary. Cindy came out onto the marsh with us. There were four curlews, two buzzards, two chiffchaffs singing antiphonally, two crows, two black-headed gulls, and two stock doves.

'Noah's ark!' said Eddie.

Then there were two oystercatchers: spring birds for us,

arriving from the coast with splendidly loud and assertive piping to claim a place for their own: black and white, carrot-beaked and with a profound dislike of silence. Not calm birds, oystercatchers: there seem to be no small disasters in their lives. Everything is an emergency, especially spring.

And from the heronry we could hear the odd throat-clearing bark and a clattering of bills. It seemed that life was continuing there as well. There was no longer any reason to doubt this. The pleasure of this realisation went astonishingly deep. I raised my bottle of beer in their direction.

'Cheers to herons!' said Eddie, who loves a good toast.

The Blue Whale

David showed me the skeleton
of a Blue Whale
the Blue Whale
eats a lot of food he said
the Blue Whale is so big
it can only live in water
I see the Blue Whale
in the sea
I feel happy
excited
to see blue blue sea
and light
shining on the Blue Whale
I see its food
krill
small like my little finger
the Blue Whale eats a lot

a lot
of little krill
the Blue Whale
lives in the ocean
all the time
men search the seas
for the Blue Whale
to take photos
films for people
like me
thank you
David Attenborough

10

BOUNCEBACKABILITY

Hurrah! Shooting Times is now following me. Or is that stalking?

Sometimes all you see is the vulnerability. So you miss the strength. The resilience, the toughness, what a football manager called bouncebackability.

Vulnerability is a fact of daily existence, and looking after is still necessary. But if you only see the vulnerability you're not seeing the whole. Two more or less simultaneous events proved this point: Eddie's ECG and heart check-up, and the return of the herons to the heronry.

They said Eddie's heart was a damn good thing: beating away like anything, doing its job just as a heart should. And that's remarkable. When he was born his heart had two bloody great holes in it. He had open-heart surgery at four months. I have the most powerful physical memory – what sports psychologists call a psychokinetic memory – of holding him across my left shoulder in the time before the op: a little flop of nothing with scarcely the energy to breathe, just enough to continue the task of Not Dying. He carries the scar right down his chest: 'my zip', he calls it. I remember the serene confidence of the medical staff: they knew the problem and they knew how to fix it, and their attitude made the process a good deal easier to deal with.

They fixed it all right. And that revealed the strength that lay behind it all: Eddie's desire to live. It wasn't that we were wrong to worry, or wrong to be fearful, but it was the strength, the resilience that dominates our memories of that time – and, for that matter, our experience of the present. It's the future that brings worry: when we concentrate more on vulnerability than on strength. Perhaps that's the right way to look at it.

We could now hear the herons not only every day but most of the time as they re-established themselves in their lofty nests a quarter of a mile away. The nests were mostly hidden from us by the budding twigs and branches: but we could see the big birds flying in, hear the barks of greeting and warning, and occasionally see a wing rise up higher than the trees.

I'm not about to get Panglossian on you, telling you that

everything is all for the best, and all we have to do is wait for the appropriate bit of strength to manifest itself. I have walked in places where rainforest once stood, I have seen oiled birds drowning, I have seen an extinct bird. But there is strength as well as vulnerability in nature, and given half a chance it will make a fight of it. The herons were back in the heronry, a life-affirming thought. And here's another: Red Nose Day was coming and Eddie had decided to go to school as Slash the guitarist. Two good things, then. Cindy found a dressing-up wig and doctored it accordingly; Joseph lent him a Slash trademark top hat, which he had once worn on stage, some years back as a homage.

Eddie looked sensational. The heronry wasn't looking so bad either.

> Heron takes flight from the dyke: reed between the lines of his beak.

No doubting the strength, the liveliness of nettles. I walked along the areas opened up by Lancelot Gobbo and his dinosaur: no bare earth to be seen. The first time he had performed this service I was convinced that the drastic treatment would let loose the ancient seed deposited years ago in the soil, and a thousand flowers would bloom. We got nettles.

Now as spring advanced, the bare earth was colonised by bright green, eager nettles, all ready and willing to sting. By the time high summer was here they would be five and six feet high: impenetrable thickets of serrated and vindictively green leaves. They seem like a smug reproach to us humans: you think you can do what you like with this landscape, but you can't. You have to do what we like. Go on! Bring out the weedkiller! You'll poison everything else, but we'll survive and we'll come back.

The reason for this eager proliferation is simple enough: nitrogen. Fertiliser runs off the agricultural land all around us. It's used to enrich the meadow grass and to grow the arable crops. Nettles love nutrient-rich places and so they come crowding in, dictating terms and frightening off any human who might want to get close.

Plantlife is an excellent charity: dedicated, as the name suggests, to the conservation of wild plants. And they don't care too much for nettles. More delicate plants are 'bullied out of existence,' said Trevor Dines, Plantlife's botanical specialist. He said that nettles were 'thuggish' and that we've been 'force-feeding the natural world on a diet of nutrient-rich junk-food'.

Which is all very well, but it seems to me that the world has had rather too much of the rhetoric of intolerance of late. On the whole I'd prefer it if the world of conservation shied away from it. We're about life, not death, are we not?

And so yes, nitrates in the system and the over-enrichment of the environment is a bad thing: eutrophication, it's called, or too much of a good thing. Many plants – many wild plants – do better on poor soil. But nettles – and cow parsley – thrive on the roadside verges in and around farmland. Plantlife reckon that this over-enrichment of our soil is a more serious problem than climate change. I won't argue. But on the other hand, I don't think the answer is the demonisation of nettles. In fact, every garden should have some. Most do, whether the gardener wants them or not. But if you have a garden, it's worth cultivating a nettle patch because they're food-plants for most of our favourite garden butterflies: small tortoiseshell, peacock, red admiral and comma. The comma, with raggedy-edged wings, is even better than a peacock at looking like a dead leaf when its wings are closed: when it opens them it reveals gorgeous wings of orange and black. Commas were heading towards extinction 100 years back: now they are – in the nicest possible way – commonplace. It would be mischievous to suggest that they have benefited from eutrophication, but it's true that some creatures thrive in conditions that are murderous to others. Who decides which creatures we encourage?

> Morning ride. The small tortoiseshell has chosen colours to go with my horse.

If you get used to a place, you respond to sudden change. Horses have the same instinct, but even more strongly. If there's a plastic bag in the hedge that wasn't there yesterday, Mia will give it a good old stare: a change could mean danger, and a prey animal must check that out.

It was this capacity for registering change or responding to a new pattern that brought me my sighting of the lions of the marsh: I was aware that the pattern had changed – and my mind leapt to its gloriously wrong conclusion.

So it was that I saw the Wrong Harrier. I had been looking at female marsh harriers all winter: and they're chunky things. I was aware of a harrier on the marsh: and suddenly it was a fire-brigade moment because it broke the pattern. It wasn't chunky at all. Nor, as I realised a second later, was it brown – and it was certainly too pale for a male marsh harrier.

I got the binoculars on it: and saw the ash-grey colour, the black wing tips, the slight, elegant build combined with the usual wavering vee-shaped harrier flight: and glory be, there was a little swash of black on the upper surface of the wing.

If you're a birdwatcher you'll have worked out that I'm describing Montagu's harrier: known to birders as a Monty. I have seen them on the Spanish Steppes, where they are almost common. It seems impossible that they can be so much like the familiar marsh harriers, and yet so unlike.

The snag was that it was still March. The equinox had just passed, and the hours of light were now outnumbering the hours of dark, and that was a fine thing. But it was still too early for a Monty: they are migrant birds and should still be making their leisurely way towards the few breeding grounds they have in this country. At this far northern edge of their global range, they're very uncommon. There's a handful of breeding pairs in Britain every year: no more than that. They nest around the Wash, and on downland in Hampshire and Dorset, but they're birds that are kept secret because of the value put on their eggs by the mad, criminal subculture of the egg-collectors.

I'd been told that a pair nested pretty close to our bit of marsh just a couple of years ago. They never showed up at our place, alas – or if they did, they chose a time when I wasn't looking. So if there was a Monty at my place, it was a significant thing. So I'd better do the proper thing and make sure the right people knew.

Carl knows all the right people and, what's more, he's as good a birder as ever looked through a pair of binoculars. His instant response was that it was too early – and if there was a Monty about in Norfolk, he'd know. He's in touch with the birdwide web, and every vibration of it reaches him. So I thought, well, fair enough. But I got in touch with the British

Trust for Ornithology anyway, and spoke to Andy Clement, not only a good egg but a very knowledgeable one.

'Thanks for telling us, but I'm sure it's not a Monty. No one has reported one. So it has to be a male hen harrier. If you've been looking at marsh harriers all winter, a male hen harrier is going to look much slighter and more elegant. And they have a dark trailing edge of the lower wing: and it's very easy at distance to see this as a black line on the upper wing.'

I thought that was letting me down very gently. So no addition to the marsh list, but it's not exactly a bad thing to see a hen harrier. He was no doubt heading north to those uplands to try and breed. At least he had the chance to refuel at my place. He was a welcome visitor – and if you happen to bump into a Monty, tell him he's also welcome here any time.

Oh, and good luck on those moors.

> Pale male hen harrier passing through. Steer clear of Sandringham, fella!

It was now clear that there were two female marsh harriers regularly flying over the marsh. The third had moved on or met some accident. I was able to work out that there were two of them for two very good reasons. The first was that one of them bore green tags on her wings, so she had been marked by the Hawk and Owl Trust as a nestling. The other was untagged. And just to make things crystal clear, I occasionally saw the two of them together. Neither acted aggressively towards the other, or if there was aggression it was very subtle. But there was a question lurking in the background: whose marsh is it anyway? And how would things work out when we got to the business end of spring?

Only two days of madness left . . . six hares making the most of them.

The gabbling and honking from the marsh made me think I was back in Africa. But this time there was a better excuse. There was no hallucination: this was a pair of Egyptian geese establishing their shared interest in a nice wet-country breeding site. I woke to this sound every day for a couple of months when I made a long stay in the Luangwa Valley, and many times since then: so it's part of me. And like many other sounds, it's capable of transporting a person through space and time.

Involuntary memory: Proust's Madeleine: the most famous biscuit in history. (Outside Italy anyway, elsewhere it's more famous than a Garibaldi.) The sound of the Gypos took me back to the Valley as surely as the sound of a spoon striking a plate took the narrator of *A La Recherche du Temps Perdu* back to a train journey made years earlier. To the narrator it sounded like the hammer striking a wheel of the carriage in which he had been sitting, years earlier, while he was struggling to open a bottle of beer: *clong!* – and he was back on the train. (It's not widely known that Proust's last words were 'I could use a cold beer.')

The sound of the Gypos was a suitable prelude to a visit from Brian. Egyptian geese are found all over Africa: they're also found unmistakably on the papyrus of Ancient Egypt. In the 19th century the British nobility took a fancy to them and introduced them on the lakes in front of their country houses. The Gypos liked it here. And like the Chinese water deer, they broke out of their stately homes and went feral: now they're a workaday part of the British list. Brian too is found all over Africa but now, in advancing age, he lives in Dorset.

One of the facts that cement many a friendship is that some conversations are hard to find. They are not the reasons for the friendship, but they're an essential aspect of it. Ralph is the only person I can talk about schooldays with. We don't, very often, but it's nice – perhaps important – to know we can. A reference to the teacher known as The Loob or to Jim Burke's illegal boots can enrich a discussion of quite different topics. There are conversations about sport I can only have with others who, like me, have spent too long on the road covering big events.

With Brian I can talk about lions. His knowledge and experience of lions is greater than my own, but we have the same fascination. Lions are always somewhere in the background of our conversations, a bit like the lions I saw – or didn't see – on the marsh. We walked around the place and revelled in the early quickening of spring. It's always nice to show the place to someone who gets it: who understands – on a relatively deep level – why we have the marsh. For him, it's nothing like the eccentric millionaire's luxury yacht; it's a thrilling adventure, it's life.

I showed him the Doris-tossed branches that were, for nearly a full second, a pair of lionesses, and we talked about the fascination of error. 'The man of genius makes no mistakes,' said Stephen Dedalus in *Ulysses*. 'His errors are volitional and are the portals of discovery.'

But for all of us, errors can be the portals of discovery. The Monty That Wasn't told me things about the two species of harriers, and the way they fit in with the world. How many errors did I make – did any of us make – between the ages of 18 and 21? As many as the grains of Libyan sand. They weren't volitional and they certainly weren't the marks of genius, but they were portals of discovery all right.

The road to even modest expertise in any subject is littered with error. Every birdwatcher knows that. And that's why I say to everyone who has a fancy for the wild world but is nervous of deeper involvement: go out there and start making mistakes. You'll be the richer for every one.

There was a moment of sadness as Brian and I were crossing the marsh. In a stand of trees on the far north-western side, a bird suddenly burst out singing. 'Chiffchaff!' I said. Not because Brian needs telling, but to make the song a shared thing. Like saying 'cheers' when you raise a glass.

'Can't hear it.'

'Really?'

'Lost it a couple of years ago, damn it. Goldcrests a couple of years before that.'

It happens. People tend to lose the upper range of their hearing as they get older. It's a thing many people hardly notice, but a birder can measure it. The high, thin trickle of golden notes that comes down from the top of a conifer – goldcrest and firecrest – are usually the first to go. And now Brian had lost chiffchaff, so he could no longer cheer or say 'yes!' when the first migrant sings in the spring. I told him that David Attenborough can no longer hear the screaming of swifts: Brian was glad that he still had that one.

Life can be pretty mean. Not just in great things – the things that bring immense courage from the most unlikely people – but in tiny things as well. Would Job have cursed God had he lost the ability to hear the song that summons the spring?

We made a short journey to the chain-ferry that crosses the Yare, a pleasant little adventure that always gives savour to a pint of beer. We found a windless corner outside the Ferry Inn and for the first time that year, away from Eddie's company, I had a drink outside. We had pints and sandwiches and watched butterflies – small tortoiseshell and red admirals – supping from the hanging baskets.

And we talked lion. Brian wrote a book called *The Marsh Lions*, the marsh in question in Kenya rather than Norfolk. He spent five years on the project and got to know the individual members of the pride as old, if not entirely trusted, friends. Lions are the only truly social cat, and they seem to make up the rules of sociability as they go along. They have nothing like the sense of order and decorum you get in a wild dog pack, in which everybody knows his place and is happier that way. The tension between the lions' love of companionship, their brilliant co-operative hunting, and their outbreaks of quarrelsomeness, are strangely compelling.

That, and the danger. We have both experienced the terror that comes from a fractional misjudgement around lions: and we both understand the savage mixture of love and fear that comes with lions. There's no species on earth I would sooner spend time with, and this is the only species that gives me nightmares.

We get it wrong about nature. We turn away from city life and look at a pleasant chunk of countryside in search of peace. We've escaped from the rat race, we've left stress behind: we've come to the place where everything is calm and still and full of peace. And the chiffchaff sings out and adds to the peace, but his song is powered by death: the death of many caterpillars and other invertebrates. We have a particular love for the sight of an expanse of water overflown in high summer by jinking, spiralling swallows: never troubling that each jink is a little death. The consumption of flying insects is needed to power the flight of the swallow, but we don't think of that when we watch swallows of a summer evening.

If you have spent time with lions – learning, not in your intellect but in your bowels, that we humans are prey just as much as the caterpillars and the aerial plankton – then you see nature as something more than the bringer of calm to harassed city-dwelling humans.

Not that I'm about to start raving about redness in tooth and claw. When people do that, they are either making a contrast with the civilised way that humans live or justifying the savage nature of human existence. Wildlife isn't one unending round of brutality. But all the same, it's important to remember that that wildlife is also wild*death*: the two held together in complex and exquisite balance.

I told Brian about the pride of lions I had seen as they were being woken up and bullied into setting off on a hunt by the alpha lioness, the ones that should have been singing 'When you're a Jet you're a Jet.' Then I waited for him to top this story. Two small tortoiseshells came together for a few seconds and then parted.

> The sheep have got onto the marsh again. Get off! You're baaaaaaed!

There are some errors you can show off about: you need to be pretty cool to make *this* mistake. I heard a redshank calling: a loud, clear, triple-note, strong stress on the first syllable.

This is a nice wader, mildly unusual round here, so I stopped in the middle of the outside chores to listen. The call was repeated once. And then again.

Ha! You can't fool me. Well, you can, but not this time. That's not a redshank, that's a song thrush: a song thrush imitating a redshank. If you arrived at this place blindfold and heard that thrush you'd know it was a thrush – and you'd also know you were in a watery landscape.

Song thrushes are not slavish or compulsive mimics, but when they hear a sound that's in their arc, they often appropriate it. A wide repertoire is a good thing: it shows that the bird is experienced as well as smart. Females are attracted to more complex songs: and males are less likely to pick a fight with a singer of complexities. The repertoire can be seriously wide: an individual can possess between 104 and 219 songs, with an average of 130, and each song consists of a phrase repeated two or three times. There is a record of a bird duplicating only a single song in a sequence of 85; another bird repeating 60 songs in a sequence of 203.

A song thrush will gather appropriate sounds and turn them into song, make them their own, in the process known to literary scholars as intertextuality. 'Immature poets borrow, mature poets steal,' T. S. Eliot once wrote. A grown-up song thrush is most certainly a mature poet, or rather, a mature musician: he will add his own interpretation, and with a certain bravura. What he takes remains his. They like to mimic some birdsongs more than they do others. Often they are mad for the song of nuthatch, with all its varied piping and whistling – but you won't hear nuthatch round the marsh because we don't have the stands of big mature trees that nuthatches like. Without nuthatches, the song thrush has no way of learning that song, so our song thrushes don't sound like the song thrushes you find in woodland. But redshank: that's another matter. You won't get a song thrush imitating a redshank in the middle of an oakwood.

So, like me, this song thrush was keeping an idiosyncratic and unreliable list of the species that come to the marsh: concentrating on the species he happened to like. Song thrushes will also chuck in human-generated sounds: whistles, tractor-reversing warnings, lawn-mowers. A few years back, they were very keen on Trimfones.

The evolutionary explanation for the extensive repertoire is about dominance and sexual attraction – intra-species competition – and it's unquestionably correct. But I often wonder if the song thrush thinks of it that way. Did Slash learn to play the guitar in order to be the alpha male and get all the prettiest girls? Or was he just lost in the music? Did he think: if I master this scale and this arpeggio and produce this piece of music, I'll be able to go to bed with the loveliest women in the world? Or did he play the music because he's a musician, and the women came as a rather acceptable bonus?

When I hear a song thrush, are we listening to a survival machine? Or a musician?

Both.

Obviously.

Stately sky-dance of marsh harriers. Excuse me, madam – is that man hovering you?

Joseph, being a musician, had bought himself a new musical instrument. Not a guitar this time, but a sitar. It had been custom-made for him and all he had to do was collect it. The snag was that it was in Kolkata. Joseph and I had been talking about doing the trip together, but then, to my complete amazement, Cindy wondered aloud about going herself. That was against all precedent. I had done all kinds of travel, most of it for work, during Eddie's lifetime, while Cindy kept the show going at home. Now she was wondering about the possibilities of reversing these apparently fixed-for-all-time roles. I was astounded. I was delighted that she was up for this adventure, though I was also a little intimidated. How would Eddie cope with so long an absence? How would I? But no matter which way you looked at it, this was clearly a good thing for all concerned. Joseph and Cindy set off on a ten-day jaunt, leaving me and Eddie on a long bachelor weekend. It was a challenge for us all, but a good one.

It wasn't as if Cindy had skimped on the preparation. In Eddie's bedroom there was a pile of clothes for every single day. On the fridge there was a list of activities: Hilary would take him to his drama group, Brigid would take him to Clinks, Joe and Cilla would take him to church. And they were gone.

So, with exemplary courage, Eddie got on with coping. We were trying to manage a solar system without the sun – and the moon, for that matter – but he faced the difficulties with immense good cheer. It was an adventure. We were all of us having an adventure. We'd be great, would we not?

We were. After supper Eddie and I brought the horses in together: he mixed the feeds, put them in the stables and then led the horses in. Especially Molly, of course. Every evening, a little after the horses were settled, there was a FaceTime call from the travellers, so he could talk sensibly to his mother and engage his brother on a different level: 'You're fat! You smell!' And we coped. We coped well. We had laughs. We had affectionate moments. We got up on time and left the house on time properly dressed. Last thing at night we would go through the bedtime rituals without anxiety or clinging. It was all going to be fine.

It was a bit like my anxiety for the heronry. It was worth being anxious about, I suppose, but it turned out to be much more all right than I dared to hope, and with remarkably little input from me.

We took walks round the marsh when we had time – Eddie had quite a schedule – and on one of those occasions we accidentally flushed a green sandpiper from one of the dykes. These are smart little waders, not noticeably green at all. They're relatively unusual, and they love freshwater. When they fly they show black with white bums, like an oversized house martin – so if a house martin flies up from a pool or a ditch, you've probably found a green sandpiper. They're also birds that show you that you've got some aspect of the management right.

A couple of years earlier a green sandpiper spent six weeks on the lower meadow. There was a giant puddle there for all that time, and he took a shine to it, feeding in concentrated bursts in the mud around the edge. If I'd been managing the place as a wetland I'd have a right to be pleased, but this was supposed to be pasture, so I had somewhat mixed feelings. It was as if the green sandpiper was mocking me for my inadequacies. Some time later we got a contractor to come in and drain it. It was not a complete success: in heavy rain the puddle comes back, admittedly only half the size, but it still makes for a pretty poor meadow.

Eddie and I also had the classic butterfly moment of early

spring: a flash of white tinged with orange. A male orange-tip butterfly is always in the most tearing hurry: like a sailor charging down the gangplank for a spot of shore leave, they are all agog for sex and drink. Butterflies compartmentalise their lives: the eating part they do as caterpillars, and they are utterly single-minded. Then they change their state of being and live in a quite different way: abandoning leaves for flowers, abandoning crawling for flight, abandoning eating for reproduction, fuelled by drafts of nectar.

Is it possible for any creature to look more unlike a butterfly than a caterpillar? How can a fat, soggy, lumpen, many-footed eating machine become an explosion of colour and airiness? These transformations have always enthralled us humans: the idea of shape-shifting gets to us on a relatively deep level. Transformation is at the heart of more or less every superhero story.

I remembered that an old friend of mine, at the age of 21, heard for the first time that caterpillars turn into butterflies. I don't know how he had managed to avoid learning such a fact during a prolonged and expensive education, but it was so, and the truth shook him.

'No wonder people love nature.'

He should have listened to Mike Heron's song with The Incredible String Band, 'Cousin Caterpillar':

My cousin has great changes coming
One day he'll wake with ... wiiiiiiings!

The pride of the peacock is the glory of god ...
a butterfly passes on Blake's wings of excess.

It was our last evening as bachelors. Cindy and Joseph would be home the following day, so great excitement. The evenings were noticeably lighter: it was close to bedtime when we set off to do the horses.

We prepared the feeds, Eddie as helpful as always, both of us in great good cheer. We had done it, had we not? Coped without disaster, coped without tears, coped without irritation. What a team, eh?

I led Mia in. Eddie led Norah in. And then at last Molly. And Molly . . .

Molly was walking all wrong. Oh God. She had lost control of her hind legs, she was staggering from side to side. She looked likely to fall any second.

'OK, Eddie, I'll lead her. You stay with her though.'

We got her back to her stable and even that felt like a triumph. I called the vet. Molly looked desperate. So did Eddie. Molly – beloved, beautiful Molly – was about to fall over, and when she did, she probably wouldn't get up again. Her eyes were full of bewilderment.

It was past Eddie's bedtime, but bugger that. 'We've got to help Molly, and the best way we can help Molly is by being strong and brave. Molly needs you. And she needs you strong. So go and talk to her and comfort her and let her know you're around.'

He knew the truth of that all right. So he made a massive, visible effort and went and stood at the door of Molly's box. She stuck her nose over the door and he talked softly to her and stroked her big face. And she got a little calmer.

Emily the vet, also making a huge effort, came in an hour or so. Holding for the vet is part of the horsey life. You put your hands on the head-collar and attempt to keep the horse relaxed and still while the vet does difficult, often uncomfortable and sometimes painful things. Molly was easier with Eddie than with anyone else in the world, and so I told him to do the holding. He did it brilliantly.

A lot of veterinary medicine is about treating everything that it could possibly be; sometimes a hard diagnosis is impossible, since horses are unable to speak. So Emily did exactly that, and gave us complex advice about the caring regimen that we needed to follow over the next few days. I could see from her manner – she was very careful about her words – that she didn't expect much. It looked as if Molly was a goner: but she hadn't gone yet. And when, under instructions, I gave her a little food, she ate it. She ate in the manner of someone quite keen on the idea of Not Dying.

Eddie went to bed in a sombre mood. We've had horses die on us before: horsey death is, alas, part of the horsey life. He would prefer Molly to be exempt, but what can you do?

I sat up for a while after Eddie had gone to sleep, amazed at the strength he had shown: at the strong, calm horsemanship, at the way the emergency had brought out qualities greater than I knew. Resilience. Toughness. Coping with a bad situation: coping with his own emotions. Knowing that it wasn't about him: and that's a huge concept to take on.

Russet, silver and sable: the marsh harrier reclaims the spring.

Next morning Molly was alive, brighter in the eye and even up for a little breakfast. Emily the vet made a second visit a few days later and was frankly amazed. Cindy and Joseph came back and there was a great reunion. The sitar was a thing of beauty. It sounded even better than it looked.

A day or so later a new bird arrived on the marsh. It seemed to be bright silver, picked out with a little chestnut and sable: but it was the pallor, the silver that dominated. It was a male marsh harrier: a bird handsome even by the elevated standards of his kind. It seemed to me the most beautiful bird I had ever seen, the most beautiful bird in the entire world.

A couple of days after this perfect moment, I saw the male again: high above the marsh. With him was one of the females. I could just hear shrill cries they made to each other in their excitement, in their joy. This was the ballet of the marsh harriers, the sky dance, in which they show off their full range of aerial skills, swooping on each other, evading each other with the neatest body-swerve, sometimes going talon-to-talon with each other and tumbling together like an enormous shuttlecock. At the high point – though not on this occasion – they pass food one to the other as they fly: the male showing that he is a good provider.

They had flown very high to perform this ballet. When you have the entire sky – the entire sky of Norfolk – for your stage, you don't have to keep close to any ground-bound observer. This was not a moment of intimacy for me: it was like being an astronomer, marvelling at two distant stars.

Spring was changing gear on us. Spring was getting serious. The maddest and most intense time of the year was almost upon us. It seemed to me that the entire marsh was vibrating like a gong, except that the sound didn't die away; rather it gained in volume: ringing with new life, the sound, already loud, still expanding and dilating in volume.

11

THE RING OF POWER

🐦 Fighting a frigid fearsome wind and winning – swallows!

The one great lunar festival of the Christian calendar of the conventional European year seems oddly incongruous: out of step with modern life. It provokes a sharp nostalgia for times we never knew, times when the moon played a genuinely significant role in day-to-day, night-to-night, month-to-month life.

The moon was full. As I got back home late after a visit to a wetland on the other side of the country, I could hear a fox baying in classic fashion, whether in response to the moon – or rather, the phenomenon of nocturnal light – or to some hidden evolution of vulpine politics, I couldn't say.

When it's a full moon, many of the terrors of the night are lessened. What a treat it must have been for our ancestors on the savannah, to be able to see the lions in the hours of greatest danger, to be able to respond to them and move away or try and frighten them off. In the centuries before street-lighting, the moon expanded the possibilities of the night: making it safe. You could find your way easily, avoid any dangerous obstacles, and you were less likely to be surprised by dangerous men. Samuel Pepys was always keen on the moon: on good moon nights he was able to live a quite different life: 'So back to my aunt's, and there supped and talked, and staid pretty late, it being dry and moonshine, and so walked home.'

Easter Day falls on the Sunday that follows the first full moon after the March equinox. That was decided at the Council of Nicaea in AD 325, and it remains the case today. So Easter dodges about from year to year, within a margin of four weeks, the length of a lunar month.

Easter marks the spring as Christmas does winter. The marsh that seemed dead in the winter was now ringing with life. It's a new beginning: but, thrillingly, only a beginning. There is more, much more to come – and perhaps, at least in human terms, the best few weeks of the year.

My father came up for a few days and great festivities were planned. On Good Friday, he accompanied me and Eddie around the marsh. Eddie's speech can be difficult to follow, and my father is pretty deaf these days, so you'd have thought communication between them would be impossible. But it isn't: showing what can be done by the power of the will.

Perhaps communication is more an act of will than anything else in life: and it's a great help if the will is shared. Communication is about accepting the fact that the other party concerned has something worth listening to. Communication is one of the things that floods over the species barrier, heedless of the scientists and philosophers who tell us it can't be done. I'm pretty good at understanding what horses are saying and how they are feeling; I'd have been seriously injured or worse had that not been the case. People who have dogs and cats are accustomed to reading their moods and their needs. That is obviously a two-way process: the animals we choose to share our lives with inevitably grow pretty good at understanding what is going on with us, and they respond accordingly.

This notion of achieving a rough-and-ready understanding of creatures of a different species in the course of our domestic lives helps us to tune in to wild creatures. I'm always prepared to go along with the idea that the animals I see have something to say and that it's worth listening to. I'm prepared to accept that they have lives full of meaning: that they aren't pre-programmed automata. The sky-dancing harriers were dancing not only under the compulsion of their genes, but also for the love of the dance and for the joy of being half of a pair. And who can't relate to that?

There is a difference between mad anthropomorphising – he understands every word I say – and accepting that dogs and marsh harriers and song thrushes have stuff in common with us. We humans also have blood and bones and brains, we also know pain and pleasure, we also seek to reproduce, we also sing and dance.

And just then a song broke out. It came from the tall clump of sallows, the one I keep thinking we should thin out or cut down. There, quite marvellously sweet, trickling down the scale with just the hint of a lilt, the song of willow warbler.

Just think of it. That little bird, singing hard from cover, had made his way from Africa, from way south of the equator, in order to get to our sallow trees and to sing on our marsh. He was tiny – not that I made any effort to set eyes on him, for I had no wish to disturb him in the fullness of his song – and yet his tiny wings, as Mrs C. F. Alexander wrote in the hymn, had carried him 7,000 miles: across the Sahara, round the rim of the Mediterranean finally to come to a small patch of marsh in Norfolk. Had he sung the same song in the same place last year? Had his father?

Good Friday is the most sombre day of the Christian year: here was one of the most exuberant moments in the year of the marsh. Either way, and no matter what you believe, it's all about the celebration of new life.

> On Easter Day the swallows are back in Aldeburgh. We're not from London, you know.

On Easter Day, showing immense courage and a reckless form of trust, we took lunch outdoors on the seafront in Aldeburgh. The weather responded with unexpected generosity: out of the wind, we found ourselves in a little suntrap and feasted with immense cheerfulness. Afterwards we took a stroll round the excellent RSPB reserve of North Warren and there, skimming the grazing marsh, we saw the year's first swallows: as good a symbol of Easter – of spring and new life – as any bunny.

They were the first European swallows I had seen since I was in Zambia. And when we all got home from our outing, there were house martins swooping low over the marsh: easily differentiated from swallows by their white bums, but clearly not green sandpipers.

Swallows and martins are closely related. They're both hirundines, both hunters of aerial plankton: flying insects and other invertebrates that fill the air. But there's no aerial plankton to speak of in the winter, so the hirundines fly south: warmer weather and plenty of food. Why don't they stay down in Africa and compete for the insects throughout the year? After all, a good few species of hirundines do exactly that. But that means the niche is a little crowded – and besides, that annual bonanza of flying food draws the hirundines in when it becomes available. Someone's got to eat the stuff. Where there's a living to be made, the wild world has a habit of finding the species to do the job. It's not true that every possible ecological niche is filled, but an awful lot of the possibilities are being worked.

The swallows' annual bonanza is about wings. Flying is very expensive in terms of energy, but very economical in terms of distance. It's harder to fly for an hour than to walk for an hour: but it's a hell of a lot easier to fly for a mile than to walk for a mile.

So if you're a bird and there's no food where you are, you can always fly off and try to find some in another place. Sometimes this happens on an unpredictable basis, like the waxwing that turned up in the meadow: this is an irruptive species, one that responds to changes on an ad hoc basis rather than one with deep-set patterns of regular journeying. The blackbirds that ate our apples in the winter had made a relatively short migration, from Scandinavia and northern Europe.

And some birds fly colossal distances. Of the world's 10,000 species of birds, around 1,800 are long-distance migrants: birds who make twice-yearly journeys between continents. Swallows, house martins and willow warblers are all in this category: and you could hold all three in your cupped hands with room to spare.

Migration is barely within the confines of our imagination, even though we can make the same journey as a swallow in less than a day. How much more extraordinary it must have seemed in the days before easyJet and duty-free and priority boarding. No wonder Gilbert White, the great 18th-century naturalist (his book about his own local patch is still in print, so there's something for an ambitious author to aim at) eventually and rather reluctantly convinced himself that swallows hibernate, spending the winter buried in the mud at the bottoms of ponds and lakes – because after all, when you first see them in spring, they're almost always flying over water, aren't they? They do so because it's a good place to find flying insects, of course.

One of the things you learn if you spend a lifetime in journalism is that the truth is always a better story than anything you can make up. And so we now understand that the swallows arrive from a 7,000-mile journey, ready to eat plenty of insects and – this being the point of the journey – to make more swallows.

Some species make even more stupendous journeys than

swallows. Arctic terns are also Antarctic terns, flying from one end of the world to the other twice a year. Bar-headed geese have been recorded at 21,460 feet as they cross the Himalayas. But perhaps the stubby-winged willow warblers are the most astonishing travellers of them all. They're not great fliers in the manner of swallows, which spend most of their waking lives on the wing. Now the willow warblers were back on the marsh, they wouldn't move much at all for many weeks: foraging in the sallows for small bits of insect life and nesting on the ground in a cunningly made little purse of grass. And yet, when the moment is right, these birds and their new-fledged offspring would set off back over those 7,000 miles, with as little fuss as I catch the 10.30 from Norwich to Liverpool Street.

How do they do it? Flying at night in flocks, keeping together by means of constantly repeated night-flight calls, using the stars and the earth's magnetic field as well as mental maps possessed by the more experienced fliers. The urge to migrate is triggered by changes in the length of days and by a great release of hormones: in some ways the birds that make the journey are quite different from the birds that live the busy warbler life at either end of the journey.

And here, even as I marvel, I must also worry. The climate is changing. Spring arrives earlier and earlier. The resident birds start singing earlier, the insects are seen on the wing much earlier. But mostly, the long-distance migrants arrive at the same time. Across the centuries their journeys have been timed to coincide with the peak abundance of food: the best possible time to feed themselves and a brood of chicks.

So if they get their timing out of kilter with the insects, they will raise fewer young. Fewer birds will make it to

Africa at the end of that long journey, and fewer still will be back on the breeding grounds. The numbers aren't adding up in quite the same way. The great global equation is no longer balanced. There is a flaw in the system like a snag in your favourite woolly: and the more you pull it, the less woolly you will have.

Always something to worry about on that damn marsh. The resilience and the weakness of the system are there to see, every single day, in one way or another.

> What a morning. The green woodpecker is laughing at his own joke.

Molly was still too poorly to go out into the field, but the monotony of box-rest could be broken by taking her out in a head-collar and allowing her to graze in hand – 'having a pick of grass', horsey people say. Naturally, this was Eddie's job.

The two of us gave her a good old groom, getting rid of implausible quantities of loose hair. By the time we had finished there was a great drift of white fur on the yard and blowing off into the field. More than once we have found birds' nests, sometimes from the many nesting boxes we have about the place: and these tend to be lined and sometimes part-constructed with Molly-hair. The idea that we are performing a service for the birds at the same time as doing one for Molly is rather pleasing.

Afterwards, Eddie led Molly down to the bottom of the garden, where the best grass was to be found: so far that spring it had not been touched by a lawnmower. Molly dropped her head gratefully and set off on that dogged crop-crop, chew-chew rhythm of grazers. I sat on a tree root while Eddie did the proper job, both horse and boy I think enjoying their time together. He was giving his horse a treat, and that's a treat that works both ways. Molly's pleasure – if not her gratitude – was very obvious.

It's a nice thing, grazing a horse in hand. You have to keep your concentration – make sure the horse doesn't step on the lead-rope, which can cause a moment of panic – but it's mostly a meditative process. And Eddie fell into thoughtfulness.

'Dad, how does a barn owl dob down?'

A barn owl dobbing down is one of the features of the place, something he has seen on many occasions. You see them flying, slowly and silently, and then all at once they vanish. They dob down: either to catch or to miss the creature they were stalking from above.

'You know what a parachute is, Eddie?'

'Yes.'

'When people jump out of aeroplanes—'

'Yes!'

He knew. Good. 'So when a barn owl is flying over the marsh, he's like the aeroplane.'

'Yes.'

'But when he sees – or hears – something he wants to catch and eat, he has to drop down out of the sky. You can see that he lifts up his wings. Now he's not like an aeroplane, he's like a parachute. And he dobs down without hurting himself and without the vole hearing.'

Munch, munch, munch. I had half an eye on Molly's front feet: the lead-rope was well clear, Eddie was doing a good job.

'Dad?'

'Yes?'

'What about vampire bats?'

'What about vampire bats?'

'Do they dob down?'

Where had he got vampire bats from? David Attenborough? Me? A book? School? A comic? I later learned that they cropped up in a schoolbook he had been reading with Cindy. So I explained: yes, common vampires, the ones that prey on mammal blood, they're very good indeed at dobbing down. They land close to the animal they've found. It might be a horse. And they scuttle across the ground to get to the horse's legs. Then they bite the legs, injecting an anaesthetic and an anti-coagulant – I missed that bit out for Eddie – and they lap up the blood like a pussycat with a saucer of milk. I left that last bit in.

'We don't have vampire bats on the marsh.'

'We don't. Which is good news for Molly. We have several different sorts of bats – at least, I think so – but they all catch flying insects.'

'Echolocation.'

'Yes indeedy – hey! Listen!'

Above our heads in the big ash tree, its edges now fuzzy with bursting buds, there was a song. Rather a special song: this was a garden warbler, another long-distance migrant, one I've heard singing on the banks of the Luangwa River in Zambia.

There is a special pleasure in picking out the song of a garden warbler. Not just because it's a lovely song, but also because it's not a blackcap. The two birds are notorious confusion species. The only way you can never be wrong is never to look; the two species look quite different. Both songs are musical and complex: the blackcap is more operatic, with big fruity notes and a taste for challenging modernist phrases. The garden warbler is more of a traditional folk-singer, like Bairbre in the pub: a song sweet and unhurried, rather rambling, and every time you think it's going to end there's another verse. And welcome it is too, but there's something gently comic about the garden warbler's reluctance to bring his work to an end.

Garden warblers drop by most years: I radioed up a message that this one was welcome to stay and breed. I hope he got the message as he looked down at a boy and his horse, and a full-grown human with half his mind on the warbler, a quarter of his mind on the lead-rope and another quarter of it thinking about life and song. One new arrival was followed by another: every day a new song ringing out around the marsh.

Who's next?

Morning ride. Swallows gallop past me on the racetrack of the sky.

There's a phenomenon among birders that you might refer to as wishful seeing. A group of hard-core birders can stare at the same distant bird and engage in a form of group hallucination, each convincing the others that certain plumage patterns really are discernible – and surely make the bird not a dunlin but a pectoral sandpiper.

Wishful hearing is probably less common – but equally thrilling for as long as it lasts. Right in the middle of the marsh, loud, clear and – unmistakable? – I heard a grasshopper warbler. This is a rare breeding bird for this country and so finding one on the marsh was something of a coup. I did a pointer-dog freeze and waited for the bird to strike up again. It's called a grasshopper warbler for the excellent reason that it sounds like a grasshopper. The song is generally described as 'reeling', not as in a dance tune but as in an angler hauling in a fish.

And then the hidden bird began to sing again: not reeling at all, but throwing in a stunning series of variations. I felt disappointment and delight at the same time: familiar bedfellows for most of us in many different contexts. This was no gropper but a sedge warbler: a bird you hear as a matter of routine in wet places in Britain. It's a fabulous singer: not terribly melodic, but with an endless capacity for variation: it's been suggested that a sedge warbler never sings the same song twice, as if for a male sedge warbler the whole spring was one song.

Every year we've been here, sedge warblers have sung out hard from cover. They've presumably bred, though I haven't got to the length of disturbing them while they're at it. They like the place and their songs are part of the great chorus of the marsh. A gropper – birders' slang for grasshopper warbler – would have been nice, but it's not about niceness. It's much more important than that.

This business of birdsong is an essential part of enjoying the marsh, enjoying the wild world. Or at least, it is to me. Learning birdsong changed my understanding of the world and its possibilities: it's not too much to say that.

It's not just that I can tell one species from another without needing to set eyes on them. That's not about niceness either. Rather, it's the sense of connection. It's like being an insider. C. S. Lewis wrote about the lure of the Inner Ring: how we long to be one of the cool people, a member of the in-crowd, part of the group that gets things done, gets looked up to, gets admired, gets copied. 'I believe that in all men's lives at certain periods, and in many men's lives at all periods between infancy and extreme old age, one of the most dominant elements is the desire to be inside the local Ring and the terror of being left outside.'

The learning of birdsong makes you a member of the Inner Ring. But it's not an Inner Ring of birders. It's more like being admitted to the Inner Ring of nature. The sedge warbler sings: you not only listen but, to some extent, you understand. An ambitious person rejoices when he comes to understand the running jokes of the Inner Ring at work. He is no longer an observer, he's a participant. The same process can happen with you and nature: and for me at least, understanding birdsong is at the heart of it.

King Solomon was said to possess a seal or ring that allowed him to command demons, and to understand the language of animals. It's one of the many ring-of-power legends, of the sort that Tolkien tapped into. Konrad Lorenz, pioneer of the study of animal behaviour, called his first book *King Solomon's Ring*. It was published in 1949 and it's still worth reading.

And sometimes when you listen to a fragment of birdsong, you feel for an instant that you have borrowed that great ring. You know the name of the bird and you know why he is singing. You can tell how committed he is to song, and sometimes how strong a performer he is. You have penetrated the babble and found an understanding.

There are many great poems about birdsong: Keats's nightingale and Shelley's skylark perhaps the most famous.

But both of these poems are not about birds or even about nature, but about the state of the poetic heart, the poetic soul. And that's nice.

But as I mentioned earlier, nature is a lot more than nice.

> Come down, house martins, come down! My house is your house!

Sedge warblers are gently declining. That doesn't allow them to stand out in the crowd, alas. That evening I heard the loud hoot of a tawny owl, and that's a call most of us can relate to. Once a tawny has established itself as half of a pair with a good territory for food and shelter, its plan is to stop there for the rest of its life.

A tawny owl's power – as I have explained to Eddie more than once – is his intimate knowledge of a very small space. He knows every hunting perch, every line of sight, every landing area. That gives him an edge. He can pick up anything unfamiliar that breaks the deeply known pattern: and whether that's a potential meal or a potential enemy, he can react accordingly, much as my horse attempts to do on a morning ride.

It follows that if you can get the wood right and keep it right, the tawny owl is in good shape. We don't own the wood near us, but it's in good shape, bar the prolonged February shooting.

We've got the marsh in good shape for the sedge warbler. But alas, I can't look after their wintering grounds south of the Sahara. I can't look after the refuelling and resting-up places on the migration route. I can't control the dangers of migration, I can't control the weather. I can't control the guns that lie in wait for many migrating birds. I can't control the changes in climate that make birds get their timing wrong and miss the peak number of insects. I can't control the chemicals we use to kill insects.

Conservation starts on your own doorstep. But only starts.

> House martins in a shivering sky. What – is spring on rewind?

The smooth progress of spring was interrupted by a sudden fall of snow. It was neither deep nor devastating: its beauties mildly inconvenient to us humans, but more serious to everything that lived out on the marsh. It never felt like a serious bit of weather, more like a bit of showing-off: see what I can do just when you least expect it. So with May almost upon us, life seemed to have been on a sort of live pause, the singing resuming with the return of the sun, even though the snow still lay on the ground.

There were now two singing willow warblers: one on the marsh, the other in the thicket between our boundary and the more distant heronry. This was new: an encouraging thing indeed, for, yes, these birds are also declining.

When any species of anything is declining, there is usually a complex suite of reasons. But usually at the top of the list there are two words: habitat destruction. We make changes to the landscape and so the species that made a living there can no longer do so. When there isn't any more habitat to move onto – and there almost always isn't – the birds fail to breed, or just die. Fact: there are 40 million fewer individual birds in Britain than there were in 1970. There is no longer the place for them, no longer space for them.

It's rather better than nice, then, to be responsible for an undestroyed bit of habitat.

Sitting out on the marsh, I could hear the clatter and the chatter from within, as heron spoke to heron from the heronry. Herons will snap their bills at each other in warning displays to protect their nests; paired birds on a nest go in for prolonged quiet rattling to each other. That's all about strengthening the pair-bond, in scientific terms: it's the sort of affectionate banter that keeps a relationship going.

I wondered, for the thousandth time, if there might be a pair or two of little egrets inside that colony: a mixed-species heronry is far from unusual. These colonial nesters are happy to have plenty of other nests around them, so long as their own site is not under threat.

> Two syllables, sweet, tentative and carrying far. You know, I really do believe that sumer is icumen in.

When did you last hear a cuckoo? I remember hearing them on Streatham Common as a boy; more recently and more than once I have gone through an entire year without hearing one at all. Here's yet another bird in steep decline.

But they love the marsh. So we wait for the first cuckoo, and naturally fret a little until we hear it. And then, six days later than last year, two great syllables rolling out across the marsh. It's a day of rejoicing: they come thick and fast at this time of year.

It's not an elaborate song, like the sedge warbler. A cuckoo isn't seeking to impress by the extent of its repertoire: it's trying to cover as much distance as possible. The way the cuckoo shouts out his presence – it's the males that say their own name – is traditionally gratifying to humans.

Sumer is icumen in!
Lhude sing cuccu!

But the cuckoo does it for far better reasons than pleasing us. He's trying to summon a female, and the further his song carries, the more likely he is to find one. As soon as he arrives he starts to sing: and all around the marsh we hear the song almost constantly for six weeks. We hear it at dawn, at dusk, in the middle of the day, and at night when there's moonshine.

It's the sound of the springtime frenzy: a great clamouring for life, for the opportunity to make more life. Of course, a cuckoo's success means failure for a pair of another species: their favourite hosts are meadow pipits, dunnocks and reed warblers, but it's not a human job to impose human notions of morality onto the wild world. There is not a lot of point in preaching sexual restraint to the bonobos (pygmy chimpanzees) or plain dealing to the chameleon or peaceful coexistence to the lion or feminism to the impala or parenting to the cuckoo. 'What a book a Devil's chaplain might write on the clumsy, wasteful, blundering, low and horridly cruel works of nature!' Darwin wrote. The creators of Monty Python wrote:

All things dull and ugly,
All creatures short and squat,
All things rude and nasty,
The Lord God made the lot.

It's perfectly possible to rejoice in the cuckoo's return and to feel distress for the dunnocks as they rear the huge bird that has already destroyed their own young. It's not even a contradiction. It's wild. It's just life.

Wildlife is not nice. Well, it is. But it's an awful lot of other things as well. Mainly it's wild. And living.

And it's doing all that out there on the marsh.

12

The Power-ballad of the Earth

Spring is back. Is that a piece of sky got loose, or is it a common blue?

A rabbit? No, damn it, it wasn't a bloody rabbit: it was a great big, gorgeous, sexy hare and in our bloody garden. I'd as soon expect a crocodile. In the five years we have been here, not a sign of a hare. Plenty on land higher up, of course, what we sometimes refer to round here as the uplands. Hares prefer dry to wet, and Norfolk is a great county for them. All East Anglia is pretty good: once in Suffolk I saw 60 of them all together, which isn't bad for an animal that's supposed to be solitary.

We see them all the time when we're out in the car doing local jobs. Cindy usually stops the vehicle, if there's not someone behind, and pulls in so we can all take a good look: the hare sometimes motionless, sometimes withdrawing at a staid, balanced canter, sometimes streaking away in a mad gallop.

One of the fascinations of hares is the shape-shifting thing: one moment they're probably a bunny and the next moment they're quite certainly nothing of the kind, jacking themselves up on those long hydraulic rams of legs and running in the manner of a deer, nothing like the hopitty-hop of a rabbit.

And now one had dropped down to the flood plain to come and join us. He seemed about three-quarters grown, making what looked like an odd decision, to come here on his own to make a living.

How beloved hares have become in recent years. You can hardly go into an art gallery in East Anglia without finding images of hares. Cindy has produced a few belters herself, preferring the fully stretched-out, cheetah-like galloping stride to the more static poses that hares go in for when they're trying to be invisible.

It's not even necessary to see them: people have told me about their special feelings for hares, and it turns out they've never seen one in their lives. Some animals attract a special form of love, not because they are cute or big or fierce, but because they have some kind of mystery that attracts us.

They seem to be an extension of our hidden wild selves: from the drastically cultivated fields of England steps forth a creature of devastating wildness, with a life spectacularly remote from our own – and yet, being a mammal, it is still one of us.

Come into our sitting room and you notice a large piece of sculpture: two boxing hares by Ann Richardson, presents for a rather big birthday. They fill the place up with their vitality, with their movement and with their stillness.

Real hares excel at the same contradiction. Movement catches the eye, so stillness is an important survival skill for many creatures. But you can acquire an eye for a hare as you travel through and round the fields of Norfolk. For as long as the crops are low to the ground, a practised observer can pick out small humps and hummocks that are not-quite-invisible hares. Sometimes they sit up like a cat on a mat, their enormous ears the highest point in the entire field, if not the entire landscape, tuning in to danger. At other times they hunker down, belly to the earth, ears flat on the back. They seem to do this almost as children hide by covering up their eyes, but there's no need to be patronising about hares. This stillness is a highly effective strategy: and when you walk too close, they will explode away from you as if they had been fired from a gun.

Here was a hare, scarcely more than a leveret, running through the gamut of these ploys. He – I'm pretty sure this was a male – was sometimes visible from the sitting room, looking nearly as lively as the sculpted hares. I saw him from the bedroom; I saw him more than once from the kitchen, when I was cooking.

Not all day and not every day: but within a few weeks he was a recognised feature of the place. He had come to stay:

or at least, to see how that played out as a general plan. He was most often seen on the higher meadow, which is more like classic hare territory, but the garden was full of grass and though there wasn't much room in it to get up to full speed, it was reasonably free of predators; presumably he was too big and too swift for the stoats. So he stayed and he nibbled and he prospered.

Already he seemed to feel safe at our place. Sometimes when he saw us, he would go into that long, flat gallop, but more often, he would jack himself up on those long levers and canter with stiff elegance to a safe distance.

Perhaps he was a pioneer: the first of a new generation of wet-loving creatures. Mad marsh hares.

A female cuckoo. And the day bubbles over.

We can't take things for granted when it comes to shared knowledge about nature. It's no longer safe to say that everybody knows what noise a cuckoo makes. Much has been made of the way the *Oxford Junior Dictionary* revised itself and chucked out words like adder, heron, kingfisher, minnow, thrush, blackberry, bluebell, bramble and poppy, and brought in blog, broadband, voicemail, attachment, database, chatroom, bullet point and cut-and-paste. There has been a spirited counter-attack in a book called *The Lost Words*, which celebrates – and attempts to make sure we keep – the words no longer relevant to the dictionary-makers.

So let me reassure you or inform you or remind you or needlessly explain to you that a cuckoo says cuckoo. That is to say, a male cuckoo seeking a mate, filling even the vast skies above a patch of Norfolk marsh with the imploring twin syllables of his name. It is a sound of hope, of long struggle, of anguish and frustration. There is a scene in Fellini's *Amarcord* when the main character's mad uncle, let out for the day, climbs to the top of a tree and shouts again and again: '*Voglio una donna!* I want a woman!'

That is precisely the strategy of the male cuckoo, exactly what the cuckoo is singing about. Will it come? Yes, it will. Maybe just by holding still. Maybe tonight . . .

Sometimes at this time of year, though very rarely, the air will be filled with a passionate cackling and bubbling: sudden, unexpected, and not at all part of the daily soundscape of the English spring. It is alien, startling, likely to make even the most avid non-birder, the most determined anti-wildlifer, pause for a moment and wonder what the hell is going on.

If you happen to have the translation, that brief and often impossibly loud burst of song is a doubly glorious thing.

You've bloody well got one!

That's what it means. It's the answer to the cuckoo's call: it's a female cuckoo that's singing back. Like the line from the Roy Harper song: 'She's the one who throws her pants at you and says, OK you're on!' It can also be heard after the female has laid an egg.

And there was the call, richly bubbling out of nowhere and changing everything in a single second for the male who had been calling and calling. That journey to West Africa and the winter in the rainforest, the return trip, the finding of the marsh, the commuting from one song-post – technically it's a stud-post – to the next, all that energy of the interminable song with its two notes, though sometimes, when extra frenzies shook him, there were almost three: cu-cuckoo! All this was worth it for the single moment of the reply, and for the few minutes that followed. (Though it's possible that the male – or another – had already got lucky and the female was delivering the result of their coming-together.)

I rushed from my desk with binoculars in hand and caught a glimpse: that hawk-like silhouette of the cuckoo. It was the female, must have been because the bubbling call was so close and her flight was sharp-winged and arrow-straight, heading in the most direct line possible – that was how I read it – towards the stud-post and the awaiting male.

That brief tryst would define both their lives: allow them, if all went well, to become ancestors, to pass on their immortal genes, to do something to bring more cuckoos into the world. Then each would fly off: the male, when the excitement had died down and desire had reasserted itself, would resume his cuckooing while the female, in a few days, would lay her egg and trust to the fostering skills of the luckless dunnocks, reed warblers or meadow pipits – dunnocks are your best bet around here.

It would be easier to listen to the cuckoo's calls for the rest of his stay, reasonably confident that he had got lucky at least once. It seemed that the year was advancing apace:

Sumer is icumen in
Lhude sing the bubbling song!

Morning ride. Has it been raining leverets on these wet fields?

Spring mixes joy, anxiety and relief, and the combination

gets more bewildering with every spring. That mixture of emotions is caught in the Ted Hughes poem about swifts:

They're back!

...

Which means the globe's still working

He makes the globe – the earth – sound like a Heath Robinson device, a crazy machine put together with a mixture of crackpot ideas and dodgy technology, likely to be thrown out of balance at any moment, but somehow still rattling along. The return of the swifts gives us at least temporary relief: the show can continue; this precarious, ill-balanced, unstable device – one that in recent years has become increasingly unsuitable for its purpose – is still spluttering and ticking unevenly onwards. Yes, we've got away with it for another year, and things are not greatly worse than they were last year.

I was pretty sure I had seen a swift a few weeks earlier, following the river course at exuberant speed. Just the one. Both the timing of the visit and the impression I got from the sighting made it unlikely to be a common swift, the sky-screamers of sun-warmed cities and countryside. It seemed a little too large, the wingbeats a little too deliberate, but then I only saw it for about 2.4 seconds. Birders will know that I am trying to claim an alpine swift here: a bird that sometimes turns up in Britain after overshooting its destination during the spring migration.

Alpine swifts always remind me of the time I was covering the Olympic Games in Athens in 2004, when most days I could see them from my bedroom window as they flew over the Olympic Stadium – the one in which Kelly Holmes performed her majestic double. And here, seen from my place in Norfolk, was just such a bird. Well, probably. Well, possibly. Birders' notebooks and conversations are full of probs and posses: terms that can reflect good observation, meaningful experience, profound knowledge, crazed ambition, wild hope and unabashed hallucination. And this was – quite definitely – a poss alpine swift. So not one for the list, but nice anyway.

But where were the swifts? Where were the common swifts, the European swifts, the real swifts, the black sickle-winged fliers that perch on air, sleep on air and even mate on air, the birds that spend the first two or even three years of their lives without ever once even perching?

I remember seeing them in uncountable numbers along the distant river: not so much a flock as a swarm, and I estimated 1,000 birds. But so far I had seen none at all. I was worried about the swifts. Damn it, I was worried about the globe.

Then, with May already in double figures, I saw two or three of them flying over an arable field at the top of the lane. I was on horseback at the time, so that made it a double-good morning, even though the swifts seemed too few and too late. Still, it was a good deal better than none. And that evening there were more, along the river. Not many. Not exactly a swarm. But anyway, swifts.

Next day I saw four or five swifts crossing the marsh.

> Morning ride. Two hares show how they have reimagined the gallop.

Sometimes I have a dream of an impossible avian plenty. It's a bit like the painting of Eden by Rubens and Breughel; Jan Breughel did most of the wildlife. In the tree the birds are so plentiful they hardly leave any room for the branches; in the water almost the entire sparkling surface is hidden: red and blue macaw, turkey, golden pheasant, goldeneye, purple gallinule, pintail, heron, mute swan; many others, more or less wing to wing in paradise as the naked human couple consider the possibilities of adding apples to their diet.

Perhaps that's where the dream-image comes from. Or perhaps Breughel was painting something common to us all: an eternal human archetype; a lost landscape of profusion, of endless variety and endless abundance; a place where the rare is commonplace.

There is a mild element of frustration in the dream: the birds are so many and so various that it's hard to identify them. I seem to be familiar with all of them yet unable to give a name to any of them. Or if to some, then not to all: not to the loveliest and the best. It's not a dream that goes anywhere: the vision fades with the dreamer still too amazed and too delighted to make a proper record: a not-unfamiliar sensation in daily life.

When I wake from such a dream there is always an element of regret, as there is in waking from any really good dream. If only I could have stayed there a little longer . . . and if only I could print from the screen of my mind the dream just gone. I can see it in general terms, but I want to see it in specific terms. Are those birds all real species, authentically remembered by my sleeping mind? Or are they mere fantasies, impossible birds, the kind painted by the sort of artist who – unlike Breughel – or Cindy for that matter – can't be bothered to do the research?

The marsh is not Eden: the shortage of swifts, the sound of the guns in the colder months, the excessive fertiliser and the million problems of the surrounding world make that quite clear. But every so often it can feel just a little like Eden: as if there were, just like the painting, rather more birds than the place can conveniently accommodate, and rather too many of them fabulous. It was May: the dawn chorus was at its height. So dawn was beckoning.

There are always good reasons for not getting up at 3.30. But I was beginning to run out of them. This was, after all, the great month of song.

> Grass snake says this place is wilder than some people might think.

On the other hand, getting up early lets you off the power-ballads. Eddie was going through a phase of breakfasting with power-ballads, and for some weeks he had adopted John Barrowman's version of 'I Am What I Am' as a kind of personal national anthem. So a pre-school morning comprised cereals with bananas and a glass of apple juice, and a chance to sing along.

> *I don't want praise!*
> *I don't want pity!*

It's no doubt a gay anthem for Barrowman. Eddie's interpretation was also deeply personal, loud and inclined to recklessness about the precise note. He ripped it out with passion and defiance. It's a song about not being like all the rest. He would sometimes explain that he had a headache 'because of Down's syndrome'. There were sad times at the end of his mostly good times at his junior school, when he was the oldest one left. When he was shorn of his friends and protectors, some of the little kids sensed his vulnerability and took advantage. And as he gets older, there is a pressing need for facing the world in the right way. He knows he is not like everybody else. He knows he is who he is.

I read the sports pages and drank a cup of tea, as Eddie went for the encore: once more with feeling.

Your life is a sham
Till you can shout out
I – am – what – I AMMMMMM!

Well, amen – and have a good day at school.

That evening we went out on the marsh to look for snakes.

This is good sport. I have laid three one-metre-square chunks of corrugated metal in various spots on the marsh, and on a warm spring evening, it's worth walking round the marsh and lifting them up, one by one. It's an old trick: it creates a warm dark space underneath and that's a lovely thing for reptiles. They don't generate their own heat; they borrow it from the sun: and the metal squares – 'tins' in the jargon – give the sun a helping hand.

I always offer Eddie the chance to lift the tins himself. So far he has always refused. It's not exactly frightening, but a successful lift does rather make you jump. And at the first tin we came to – the one by the old willow stump that Doris knocked in half – we got lucky.

I lifted the tin, revealing a square of bare earth, dried vegetation, white roots, and small holes dug by rodents, and right in the middle, making a wonderful pretzel of himself, all dark green with a shining yellow collar, a grass snake. No more than two feet long, and about two fingers thick. He held perfectly still while we both got a good look, and then suddenly took himself off in that uncanny fashion of snakes.

Grass snakes love wet country. It's hard to see them; all snakes are experts at hiding. Their strategy for life, starting with their body shape and their super-economical leglessness, is designed for easy concealment. The tins are a technique for bringing them into human awareness.

I remembered a jaunt Eddie and I made on the canoe on the Waveney, when I sighted a grass snake swimming across the river, and performed a handbrake turn – try it some time in a laden Canadian canoe – in order to give Eddie, in the front end, a grandstand view. It was a triumph: one of those unforgettable wildlife moments.

We replaced the tin, drew a blank at the other two and then went to sit on the benches for a while. The singing was comparatively quiet for the time of year. The afternoon was drawing on but there were still a few hours of daylight left. Most species seemed to have agreed on a break before the final frenzy of dusk. But not the whitethroats.

Whitethroats are yet another species of warbler. As with Eddie, ostentatious musicality is not really their thing. The song is usually described as 'scratchy' as if they had a sore rather than a white throat. And there were two males singing antiphonally from adjoining clumps of bramble. It wasn't exactly a musical treat but, well, they are what they are. I enjoyed this duet very much. Without the brambles there would have been no whitethroats: no matter what some people and some other species would prefer me to do with the marsh, the whitethroats thought I was fab.

We went back to the house in good humour. We are what we are . . .

Morning ride. Must update the protective clothing. Between these May hedges I need ear-defenders.

I left the house a little after 3.30, dressed for an unusually cold winter's day. That's experience: if you're going to sit still for a long time of a May morning, you need more layers than you'd believe possible. I wore enough to stop a bullet. I carried my binoculars, not that I expected to be using them very much, and a flask of tea. It was light enough to see without a torch, just about: the moon helped. The idea is to be there for the first moment of song in the day, but that's impossible, you'd have to start the day before. The singing never really stops in early May: there's always some bird or other ready to make himself heard even in the quiet of the night. It's like a football hooligan during a two-minute silence: he can't help himself, he just has to break the quiet with the sound of his own voice.

The Cetti's warbler is particularly keen on this: a sudden exclamation to the darkness, reminding others of his kind that he's still there and in control. I heard the first Cetti's shout as I opened the gate to the marsh and walked towards the benches and the low table. I poured tea: hot, milkless and sugarless Rooibos tea, a habit I had picked up in Africa. It seemed to act as a link between the wild world and my half-woken self, reminding me that I was supposed to be enjoying myself. The Cetti's helped: every time I have done this dawn vigil – every May since we have lived here – the Cetti's has called me to order, reminding me why I had abandoned sleep and embraced mild discomfort.

I had the world to myself. Or rather, the opportunity to share the world with warblers. There was no sound of the busyness of the farm next door, no urgent anxiety calls from reversing tractors, no sound of engines. Occasionally, from the road, I could hear an early-morning – or late-night – vehicle, but that only emphasised the isolation: the thrilling loneliness of night.

A sedge warbler started up: at first tentatively, but he soon found his stride and dropped into the pattern of endless unenigmatic variations. I heard the splashing of a female mallard swimming down the dyke: sighting me – the light was already getting better – she took the wing with a quack of alarm. There was stirring in the heronry: I could hear the sound of movement, wing-flapping and bill-clattering. No singing for herons, the only territory they need to defend is a beak's length from the nest: but an urgent day of fishing and feeding the young lay ahead.

Then, oddly, I heard the voice of a peacock. There are plenty of feral peacocks in Britain: when I lived in Suffolk they were regular visitors to our place, and their voices echoed round the village. But they don't move about a lot: they're not great fliers and they generally find a way of foraging – and getting fed by humans – wherever they happen to be. But here, perhaps for the first time, I heard a peacock: not actually on the marsh but away towards the farm. And I never heard him again.

One of the strange things that happen in the course of this annual dawn vigil is that sight becomes secondary: and that's a rare thing for us humans. It seemed that every noise I made myself was the most terrible din: the ring of the flask as I poured myself another cup and banged it clumsily against the table; the rustle of my waterproof as I reached for my cup and sipped.

A bird cleared its throat and I instantly knew it was a blackbird. A second or so later it went into full song. A little later exactly the same thing happened with a cuckoo: I knew the birds even before they had revealed themselves through song. I had recognised them when I couldn't see them and when I couldn't properly hear them either. There was a deeply satisfying sense, partly of my own brilliance but far more of being in tune: of being at one with the birds and with the place in which they had their being. I, silent, felt perfectly in tune: a member of the Inner Ring.

It was then that the single most electrifying moment of the spring took place. Perhaps of the year. A new voice joined the chorus: an insistent, drilling, insect-like call. But no insect. Might you call that reeling? Surely I had a grasshopper warbler here, surely this one wasn't going to turn into a sedge warbler: the song was too persistent, too committed, showing no sign whatsoever of breaking into the thousand variations. It sang again and then again, and I listened with a smile of delighted incredulity. Was it really a gropper, though? It sounded a little light, insufficiently persistent. Doesn't that make it a . . .

Then a willow warbler struck up from the clump of sallows, and a wren, from right behind me, drowned out everything else out with a startlingly loud and explosive burst of song, with a trill so fulsome you'd have thought he'd burst his small body in half.

After that (was this irony?) I saw something – actually saw something using my eyes rather than my ears – and it was a bird you always hear and hardly ever see. It was a tawny owl, crossing the open ground between the heronry wood and the wood beyond the edge of the marsh, where I believed the buzzards were nesting. Almost everything else in the course of the vigil came from hearing: but here was a notoriously invisible bird in plain sight.

I sat there for an hour and a half, by which time it was light and the early-morning singing had died down, though the reeling bird was still reeling. The whitethroats waited till I was on the move before they began: slugaroosts. There was a chiffchaff and a chaffinch singing in the garden.

It was like my dream of plenty, but the images were all in sound. Not a vision of beauty: and being such sight-crazy creatures we don't really have a word widely used for a thing heard. All the same, it really was an *audition* of beauty. The power-ballad of the earth.

Morning ride. Do the young hares flee for fear of the horse or for the joy of running?

Not a gropper. I looked it up before going to bed that morning, listened to a recording. It was a Savi's warbler, beyond any possibility of doubt. And that was a pretty extraordinary thing. There aren't many of them around in this country. This is the extreme northerly tip of their range: in any one year, you may get a pair or two breeding in Britain. Seldom more. Some have predicted that will be the next Cetti's: marching forwards confidently, northwards into the climate-changing world. But it's been a long threatening and still there are no more of them. Perhaps they are climate-change sceptics: Trump's warbler. (Paolo Savi was professor of Natural Sciences at Pisa in the 19th century.)

I contacted Carl, my own personal rarities committee, and he gave the bird a cheery thumbs-up. That same week, he told me, Savi's warblers had also turned up at Titchwell and at Minsmere. These two are the top two names in wetland nature reserves: not bad company to be keeping. It was as if a great musician had decided on a British tour: Covent Garden, the Royal Albert Hall – and our back garden.

I never heard the bird again, so he didn't stay and breed. That would have been a coup. His brief presence was a thrill, perhaps even a validation. But there are more important things in life – in wildlife – than thrills. The fact is that other less glamorous species were actually breeding out there on the marsh. Species in decline, species to worry about: and they were hard at it all around us. That new entry, that new tick on the Marsh List was all very fine and dandy, but the willow warblers were actually out there making more willow warblers, and that's what it's all about.

Glamour is all very well. But what matters more: the famous person you know slightly and sometimes have a pleasant evening with, or the lifetime friend who'd drop everything the moment you asked?

Sundowner, slow, very slow. Over the marsh, swift. Very swift.

Apple juice for Eddie. Beer for me. And a flask of coffee for Cindy: she was coming with us, so hurrah. Cheese and tomato sandwich, hot baked beans in a jar, yoghurt and fruit in another jar. So let's go.

There are times when it seems that the earth itself is in a good temper. There seemed to be nothing to worry about: not where the next bit of work was coming from, not what Eddie would do next year and every year after, not the declining numbers of cuckoos and willow warblers, not the changing climate, not the ecological holocaust. Everything seemed suspended. It was a bit like the interval in one of those absolutely enormous plays: *A Midsummer Night's Dream* or *King Lear*; I'm not sure if we're taking part in a tragedy or a comedy – no doubt both – but it was a pause in the action, a moment when you could have a beer and a small joke and an affectionate word before going back to the terrible questions and the still more terrifying certainties of the main action.

It was a benign world. For half an hour at least.

So we sat and we talked and we sat and we listened and we sat and we watched: over there, between the benches and the lone oak, the one that stands a couple of hundred yards beyond the boundary dyke, I saw the pale male marsh harrier drop down. I waited for him to rise again, and here's a thing: he didn't. There's more than one possible explanation for this, of course – but one is that there might be a nest there. And the air was full of seed: trees – was it the sallows? – had vented thousands upon thousands of these tiny weightless fluffy seeds, one of which might land and sprout and become a tree and produce such seeds in its turn. Everything you saw or heard was fecund.

I had a sudden memory of my fourth term at university. It wasn't spring, it was autumn of course: but in student terms that's a time of new beginnings. And it was as if the whole world – or at least every single person I knew at the time – had gone mad. It was as if the air was full of seeds and songs. Everybody who was partnered up became unpartnered and more or less immediately repartnered. Those who had slept alone for a year were now half of a pair. Virgins became as rare as Savi's warblers. People you'd never suspect of such enormities had girlfriends and boyfriends, and then busily set about being unfaithful to them. 'Imagine taking an X-ray photograph of this place at midnight,' said my own new girlfriend as we lay together in her hall of residence. 'Every single room in the entire place, there's people hard at it.'

'Lovely copulation bliss on bliss,' I replied, for I was reading Blake at the time.

The marsh was like that: like all of student Clifton in October 1971. The difference was that here, after the partnering, there would be babies.

I gave a sudden shout and a point, and two more heads turned and we all of us saw, low to the water and moving with immense purpose, a perfect vision – not an audition – of electric blue.

'Kingfisher!' said Eddie.

One of the lost words, as it happens. But not here.

And down the dyke a little way, at the end of a small, surprisingly deep tunnel, was there perhaps a collection of tiny white eggs, or little fluffy not-yet-blue kingfisherettes?

It seemed then that the wave of spring had reached its maximum height up the beach. What was left was the living, the rearing, the hard yards of the wild year. The bits we missed out in Clifton.

The year was turning.

Again.

13

I AM NOT A NUMBER

Morning ride. Overnight rain has turned the lane into a green tunnel.

I have an awful lot of field guides. Mostly birds, of course, but the inadvertent collection covers a good few other taxa as well. They all help a reader – or rather a looker, a peruser, a thumber-through – to work out what species have been seen. They are – necessarily – so intent on this often difficult and complex task that you'd think species was the end of the matter. Once you've diagnosed the species: well, job done, all questions answered. But when you spend a lot of time looking – and listening – in a single place, you begin to realise that species is the beginning, not the end. It's the answer to question one, that's all: and there are a million more. Just about every field guide I possess reads as if every bird of the same species looks the same, sounds the same, acts the same, thinks the same, is the same. Sometimes the males and females look different, sometimes the immatures look different, but crack the code and you've finished the job. You can't blame the field guides for this: they're only doing their job. It's up to the peruser to take the next step.

Pale and Dusky were so dissimilar you might almost think they were different species. And yet they were both male marsh harriers, both using the marsh, though at different times. I had already seen them many times, and rejoiced in the fact that I was able to tell them apart. Not *a* marsh harrier but *that* marsh harrier. I remembered *The Prisoner*, the mad thriller series of the 1960s, in which every week Patrick McGoohan resisted the latest attempt to destroy his individuality. 'I am number two. You are number six.'

'I am not a number! I'm a free man!'

I have ridden a lot of horses in my life: and every single time I sit on a horse for the first time, I am taken aback at the way this one is different from every other horse I have ever ridden. There are a thousand differences between every individual, sometimes subtle, sometimes very much not so. I ride Mia in a way I would not have ridden her predecessors because I know her as an individual. She has her idiosyncrasies of nature and I am accustomed to them, so without even thinking about the matter, I make small pre-emptive corrections before the error has been committed.

It had rained heavily in the night and as Mia and I set off down the lane, the tops of the hedge-trees had bowed over, creating a long green tunnel that was pleasant to ride through. I held the reins by the buckle, because I knew I could take such a liberty, and we set off in great good humour, coming back an hour later better pleased with the world than when we set off. And from the big oak just by our gate as I rode in, I heard a song that had me baffled for a moment. So we paused for a better listen: this song was impossibly rich and flowing and still I was struggling to put a name to it.

Then the song shifted gear. It was still loud and fulsome and exceptionally well phrased, but comparatively commonplace. And I knew it all right: it was a blackcap. But an especially good one. This is yet another species of warbler, sometimes called the northern nightingale: some claim the song is as good as a nightingale's; and the blackcap's range takes it deep into Scotland, while nightingales are restricted to Southern England and East Anglia.

It is a fabulous singer: and here was one with a different song. So perhaps the aim of everyone who watches birds, who looks at and listens to any form of wildlife, is to be able to recognise each individual, and celebrate not just its similarities with the rest of its species but also the things that make that individual animal unique. To get even slightly closer to that impossible ideal is a major change in perception.

I could hear the individuality in that song. It seemed to me a cut above normal blackcap song, which is lovely enough. But then I wondered how a female blackcap would take it. Did the song say to her: here is a super-blackcap, an *übermönchsgrasmücke*? Or did she say, this is wrong, this is an aberrant bird – a freak, yuck – and I wouldn't touch him with a bargepole?

Either way, it's her choice. The males sing the song of themselves and the females decide what is pleasing: what marks the sort of blackcap they choose to sire their next brood. It's a notion called sexual selection: another form of competition within a species. It's a competition that comes down to the individual. Darwin came up with the notion and the Victorians hated it. That's because its basic premise was that the decision on what matters in a species is decided by the females. But it's the right answer.

When I was writing a book about birdsong (*Birdwatching With Your Eyes Closed*) the publishers were much taken with the idea of inventing a Shazam for birdsong: that is to say, an app on your phone that will listen to a bit of birdsong and at once tell you what species you're listening to. I wasn't a great help at the meeting. There was a lot of exciting talk about algorithms, but I kept saying that it wouldn't work because it couldn't work. Birds are not the same as records. They're individuals and they vary. Sometimes considerably, as that blackcap showed. Your app is never going to be able to sample the full range of individuality that lies within a species. Alas, I was proved right, and the idea was dropped.

These days, there are products that make a decent fist of the job. The reviews I have read say that they can't get the answer right every time, though they can sometimes be helpful. But I think that on the whole it's better to learn the songs for yourself. Our brains are better than machines when it comes to coping with the notion of individuality.

Morning ride. I greet the Chinese water deer in my best Cantonese: Josun!

We can easily recognise animals as individuals, rather than representatives of a species, when we live with them. If you have two cats in the house, you know that one always sleeps on top of the fridge and merely tolerates human company while the other prefers the laundry basket and actively seeks people out. Recognising and understanding individuals is infinitely harder with free-living wild animals, and doing so has been the great achievement of many of the great ethologists – students of animal behaviour – of recent years: Cynthia Moss with elephants, Jane Goodall with chimpanzees, George Schaller with lions and many others subsequently. People have done it with gorillas, bonobos, whales and dolphins, hyenas, wild dogs and many, many others. And it all begins with the identification of individuals.

So let us allow ourselves to be sidetracked by the magnanimous wolf; I read about him in Carl Safina's *Beyond Words*. The wolf was called Twenty-one. (Some researchers give names; others think that's frightfully unscientific.) Twenty-one had a personality that effortlessly disposed of all attempts to dehumanise – or delupinise – him.

Wolves fight. They fight seriously and bitterly and often to the death. But not Twenty-one. He was so great a fighter that he never once lost – and never once killed a defeated opponent. All right, let's be careful of casual use of big words. Let's just say that if Twenty-one had been a human, we would admire him for his magnanimity, for his mercy.

Safina took this a courageous step onwards: 'When a

human releases a vanquished opponent rather than killing them, in the eyes of onlookers the vanquished still loses status but the victor seems all the more impressive. You can't be magnanimous unless you've won … And if you show mercy, your lack of fear shows tremendous confidence. Onlookers might feel it would be desirable to follow such a person, so strong yet inclined toward forbearance.'

The ability to recognise wild creatures as individuals is not restricted to people who fill days and years watching the same small group. If you regularly visit the same place, you add a deeper layer of satisfaction with the knowledge that you are often encountering an animal you've come across before, maybe many times. I didn't know if the willow warbler singing in the sallows was the same one that had sung there last year: but I did know that it was more or less certainly the same bird I was hearing every day.

And when I saw those two harriers – always separately, never two at once – I knew them as individuals. Dusky was most often on the right towards the buzzard wood; Pale more often to the left, nearer the heronry. I hadn't seen either of the females for a few weeks: I hoped very much they were both hunkered down on eggs.

This is all pretty rough-and-ready stuff, of course: not exactly good scientific data, meticulously recorded. I lack the skills and the temperament required to acquire and collate such data. Like those harriers, I too am an individual. Like the blackcap, I sing the song of myself and hope for a favourable audience – and as with the blackcap, it's not me who decides whether my song is any good.

I am what I am, in other words. I don't sing the song with the same level of defiance as Eddie, but then I don't need too. People get the same wrong idea about individuality with Down's syndrome. We have often been told, especially when Eddie was younger, 'they're very loving, aren't they?' The whole 'they' thing became one of our running jokes: the idea that all people with Down's syndrome are pretty much the same. But Eddie is no more a number than the Prisoner.

A blazing little sun instead of a head.
Yellowhammer morning.

I saw Pale twisting and tumbling in the air, taking advantage of a sprightly wind, and wondered why he was doing it. There seemed to be no food-gathering going on. I wondered: was he doing this to amuse himself? After all, if you could fly like that, you'd twist and dive, plummet towards the ground and then with a single twist of the wing, catch the gust and rise up again as if you were in an express lift. I had been out on my horse, playing the same sort of game, enjoying the day, putting my skills through a mild workout and savouring my own mild success. It's called fun, and surely it's not restricted to humans. Mind you, I never saw Dusky behaving in a playful way: was that a point of difference between them? Or did I just never catch Dusky at it? And if it was a difference, which was the more desirable bird: Playful Pale or Sobersides Dusky?

I don't suppose I'll ever know, but even asking the question took me closer to the birds and the place where they lived, the place where they expressed their individual natures.

An hour or so later I took a turn round the marsh by myself and sat on the bench, as the marsh did its stuff all about me. That same willow warbler was singing his same excellent song. And then damn me, but Pale came heading right at me. He was going flat out, which for a marsh harrier isn't all that fast, but in straight and level flight – in a rhythm of half a dozen powerful wingbeats followed by a power-glide – he was a pretty impressive sight. I kept dead still, of course, and he carried on coming towards me about eight feet off the ground: perhaps straight over the bench was a favourite route of his. The bench might even have been a favourite landmark. And then suddenly he saw me and veered hard left along the dyke. If I'd made one of the great goalkeeper-dives of my youth I might just have tipped his tail over the bar, he was that close. The only time I'd been closer to a wild marsh harrier I was holding him in my hand; that was when I spent a day with the Hawk and Owl Trust and they put wing-tags on a nest of chicks.

It's all right! I'm on your side! You don't have to worry about me. You don't have to go charging off up the dyke just because I'm here. But there's no convincing some birds. Off he went, wings held in that perfect dihedral: a marsh harrier and his marsh. All his, so long as Dusky wasn't around. This bit of marsh seemed to be a bit of territory in common to both; certainly it wasn't aggressively defended by either. Who did it belong to, then? Pale? Dusky? Neither? Both? The one sure thing was that it wasn't mine . . . and from all this, I gathered that the pickings must be pretty good round here. Well-filled stomachs are a good first step towards world peace.

There were two things vexing me in this fine and song-filled May. The first was the state of my knee; the second, the shortage of swifts. Neither the globe nor my body was working quite as it should. The offending joint had been scanned and diagnosed: I had a problem with the cartilage and it was to be operated on in a few weeks. I was looking forward to walking without inconvenience; I was not looking forward to the stuff I had to go through first.

And there just weren't enough swifts about. There didn't seem to be any of the screaming parties: those joyous gatherings of young, unpaired swifts, yet to take on the responsibilities of domestic life, racing each other and screaming at the tops of their voices as if in the advanced throes of Beatlemania. Shame David Attenborough can no longer hear them. Joseph and I used to reserve one evening every year, back when we lived in Suffolk, and take a bench outside the pub and watch the swifts from head on, screaming their way down the village street, using the houses on either side to delineate their racetrack, 20 and 30 at a time.

Not hard to see how this bird got its name. One swift was timed at 69.3 mph in a screaming party, and the average speed of the pack is around 50 mph. They can't sustain their top speed – as some ducks, geese and waders can – but when it comes to a sprint, swifts are champions.

And I was seeing very few of them. I had hoped this would change as the spring advanced. And certainly things were a little better: I saw a mixed flock of 30 swifts and martins flying over the marsh. But it really wasn't good enough.

The love of wild things has long been driven by the sense of loss. Wilderness was once something to be avoided: fearful, dangerous, desperate, deplorable. When Omar Khayyám finds that Wilderness was Paradise enow, he is paying the highest possible compliment to Thou – who you will remember was hypothetically beside the poet, singing in the wilderness. Why, if she could make even wilderness agreeable, she really must be something.

It was only when we began to destroy wilderness that we began to see its point. It was only once the Lake District became accessible that Wordsworth was able to get there and celebrate it. He was able to appreciate daffodils and rocks and stones and trees because he had seen the beginnings of industry and the increasing pace of urbanisation. The Romantic Movement discovered nature as a thing to love for itself – wild nature, wilderness, wildness and wet – because they were already deeply familiar with its dark, satanic opposite.

That process has been continuing ever since, and at ever-increasing pace. You can read the journals of birdwatchers from just a few decades back and hear them grumbling that there was nothing to be seen that day but a flock of 5,000 turtle doves – a bird now in danger of extinction in Britain. These days seeing a single turtle dove is a triumph, and one you share with your local wildlife trust, to make sure they are aware. Talk of 5,000 is the stuff of fantasy.

Conservationists talk of 'the shifting baseline', which sounds like one of Bach's musical techniques. It refers to the change in expectations: the new normal. These days it's normal to see a few lapwings, abnormal and remarkable to see large flocks. Ditto curlews, as already discussed. Conservation was invented from the same sense of loss: and yet few of the wonderful and highly motivated young conservationists at work today have seen a few thousand lapwings all together. For them, a few dozen lapwings here and there on arable land and a few more in a nature-reserve hotspot is what nature is: the ideal we should try and conserve at all costs.

Some say a sense of loss is part of the human condition: loss of childhood, loss of mother–child bond, loss of innocence, loss of youth. And yes, an inevitable loss of mobility, I could relate to that all right. We must all cope with those things, of course, so perhaps it's not loss that defines us but the way we cope with loss.

Loving nature is about coping with loss. It's about plenty of others things too, of course, and many of them are gloriously uplifting. But the glory is always underpinned by the sense of loss.

I will never keep goal again. I still ride, but those demented joys of cross-country jumping events lie in the past. When I dismounted I made sure I landed on my left foot, because my right knee wasn't able to hold me.

There were herons back in the heronry. There were harriers – once extinct as breeding birds in this country, remember – cruising the marsh. But there were very few swifts screaming overhead and, as yet, no swallows building nests in the barn.

> The bath-toy ducklings chase their mother up the dyke on over-wound elastic.

Most weekends Eddie and I bake a cake. There was a time when we made fancy yeast-leavened fruit loaves with enriched dough, and they were immensely satisfying, but these days Eddie is gluten-free. This has had the most excellent effect on his digestion, but an initially depressing effect on our baking. A number of cakes fell apart or had the texture of sand. But we got better and now we have a series of recipes that give excellent results. They tend to be simpler than the ones we made in our glory days, but that's because these days Eddie does the weighing and measuring and mixing and cooking himself, with me offering shrewd advice from a distance, as I do when he is lunging a horse. Cakes have been one of our adventures.

One of the routine favourites is marmalade cake, sometimes of course referred to as the Paddington Cake. It uses getting on for a full pot of marmalade, and if you choose a good one with plenty of bitterness, the result is a cake you could introduce anywhere.

We have all enjoyed the two Paddington films, Eddie and I especially. I was a great fan of the Paddington books in younger days.

'Are you Russian?'

'Well, I am in a bit of a hurry,' said Paddington.

Two words – 'said Paddington' – make the joke funny.

The films took the ideas behind the books and made them explicit. In the closing sequence of the first film, Paddington is writing home to his Aunt Lucy back in Peru: 'Mrs Brown says that in London everyone is different. But that means anyone can fit in. I think she must be right because, although I don't look like anyone else, I really do feel at home. I will never be like other people but that's all right . . .'

The film makes a nice pair with *Paddington 2*, which came out the following year. The first film is about the importance of welcoming strangers, and the way Paddington's arrival makes the Brown family a great deal happier. The second film is about the benefits Paddington brings to every community he finds himself in: his kindness, openness, good manners and willingness to find the good in everyone make those around him better and happier people.

I shan't labour the point, but Eddie is not a drag on the community he lives in. He gives people the gift of being able to help someone who needs it, and that's a rich thing for starters. He also brings his own kind of kindness.

Down's syndrome is getting rarer because we're better at predicting it. It's widely seen as a bad thing, and most pregnancies with certain, probable and even possible Down's syndrome are terminated. People must make their own decisions on such matters, but they should make them from a full knowledge: understanding that people who have Down's syndrome can be a positive asset to the world in which they live.

Marmalade cake tends to bring such thoughts into my head, and besides, Eddie and I had decided to take tea and a couple of slices onto the marsh.

Some would see the marsh as a wasteland, one that should be better put to more productive use – meat production, for example. And some see a person with Down's syndrome as a mistake that should have been pre-empted.

Eddie and I sat on the marsh and ate our cake and drank our tea and listened to the clatter and the chatter from the heronry. Their own parental responsibilities were at a peak: the young ones were getting big, and we could occasionally glimpse movement through the branches, as a young bird exercised wings that had yet to fly. Young herons exercise with great enthusiasm in preparation for their maiden flight. A heron is a big bird, which means that flight is a big deal. A little scrap of a thing like a wren, a bundle of feathers that weighs no more than a pound coin or two, they're so small and light that being in the air or on the ground isn't that much of a transition for them. By the time the young have fluttered around the nest area, usually inside a hedge or a clump of brambles, they are beginning to know what they're doing. It's like learning to ride a bike in the back garden. But for a heron, it's like taking your first drive in a ten-ton truck. Harder: they must not only descend from the top of a tall tree, they must also land their not inconsiderable bodies without causing themselves damage. So warming and developing those powerful wing muscles is essential.

> Spring is back. To celebrate, the hedges throw singing whitethroats at the sky.

The marsh harriers were also getting on with the job of parenting. Dusky flew past my hut – where the blob bush once stood – and naturally I suspended the current piece of work to watch him. He was about ten feet above the marsh, the classic marsh harrier height for hunting. For harrying, or for harrowing: providing a harrowing experience for anything he might find.

The *OED*'s first definition of harrier is 'one who harries, ravages or lays waste'. And there was Dusky, harrying for his life and for the life of his chicks, and all of it above our own little Waste Land. He continued that exploratory flight, slow and meticulous and of course deadly. He reminded me of the sort of screen villain who drawls, and takes things ever so slowly, never in a hurry, savouring the nuances of power. For the harrier, this is an efficient form of hunting that has developed over countless generations: those aerofoil surfaces, that easy dihedral, that comfortable rhythm of flap-flap-glide. That's his power. Buzzards hunt one way, kestrels another, sparrowhawks another and peregrines yet another: harriers have developed their own method across the generations and it works.

And then, without drama, he was gone: interrupting the rhythm of the wavering flight to drop with silent purpose – like a parachute, Eddie – down into the reeds. He stayed down, too: he'd got something. I missed the moment when he rose again: but no doubt it was a good day for his nestlings. The year was once again turning: the establishment of territories and pairs was no longer the priority for many species; it was about hatching eggs and raising young.

It's thrilling and occasionally problematic to have responsibility for eight wild acres of Norfolk. What must it be like to be responsible for 10,000? The Earl of Leicester owns land that is designated a National Nature Reserve. It was formerly managed by the statutory body – that is to say, representing us – that was then called English Nature and is now Natural England. These days the Earl – 'Hello, I'm Tom' – has responsibility for it himself, answering to Natural England. I had paid a visit there (to write a piece for a magazine) a few weeks earlier, and he very kindly invited me back to witness the wonders of late May.

Sarah Henderson, his conservation manager, met me, and we went by Land Rover to a small patch of wet woodland that lies down in a dip. The trees stand with their feet in water and they are breathtakingly full of birds and their nests: cormorants, herons and little egrets. And among them a dozen nests of spoonbills: the only place in Britain where they breed.

There's always something to watch in a big nesting colony:

the toing and froing of the parents, the squabbling with neighbours, the repairing of nests, the feeding of young, the young at various ages, changing from fluffy dinosaurs to sleek creatures ready for the air. We saw a spoonbill make its maiden flight: a thrilling mixture of over-confidence and self-doubt, like every other teenager. It landed staggering but without a crash: a small triumph. These young birds have much smaller beaks than their parents: birders refer to them as 'teaspoons'.

Land management – like anything else in the world – is about deciding what matters. What really matters. And quite often, that's not money or even power. And what was once considered waste is sometimes the most valuable thing of all.

Morning chores. In the stables the swallows outnumber the horses.

And then one evening, as I brought the horses in, a cry of alarm from Norah's box. And from it there exploded a swallow, closely followed by another swallow. I stood for a moment, watching them sketch circles and spirals over the meadow, and felt a sweet relief flowing through me. A late start, but at least a start. They were birds in a hurry. I could almost hear them say, with Withnail in the film: 'I'm making time.' They would need to.

'May's nearly at an end, Eddie. Do you know what happens next?'

'No.'

'June. And do you remember what June means?'

'Wild? Wild June.'

'June means 30 Days Wild.'

'30 Days Wild!'

He remembered that all right. We've done it the past three or four years, ever since the Wildlife Trusts came up with the idea. Simple, like all great ideas: do something wild every single day in June.

And as we've done before, we would both blog every day we could: putting the stuff up on my own website, with links to and from the Wildlife Trusts website.

An imposed discipline or structure like this is a good thing, a helpful thing. It would get us out there: and it would get us both telling the world about what we'd been up to.

And what would that be? It's possible that the odd beer, apple juice and picnic on the marsh might come into it.

14

RUNNIN' WILD

Morning ride. Enough small tortoiseshells to make a flying Galapagos giant.

There was a southern marsh orchid by the pond in the middle of the garden. Even I knew that: even I could recognise that dense cone of tiny purple florets. Beside the orchid, the flag irises were waving their yellow banners, so Cindy, Eddie and I took some refreshment, sitting on the grass around the pond. Easy conversation was not a straightforward business because I kept staring at the pond through binoculars, trying to will every dragonfly to possess green eyes.

I have a pair of ultra-close-focusing binoculars and they're wonderful for looking at detail. At least, they are once you've got something lined up. But focus is both critical and elusive with these binoculars. When you're trying to get a bead on a fast-moving creature like a dragonfly, you are likely to get frustrated and to irritate those around you by your inattention.

There's a species of dragonfly called the Norfolk hawker. They're not much found outside the county but, by rights, they should turn up here. In fact, I'd be prepared to bet a fair amount of money that they already do; it's just that I'm such a poor observer, especially when it comes to dragonflies.

Years ago, I wrote a book called *How to be a Bad Birdwatcher*, and I am a bad birdwatcher to this day. But I think I can claim without undue modesty that I'm a *good* bad birdwatcher. Perhaps decent non-League standard, though when it comes to song and call I might just scrape into the old Fourth Division.

I am a pretty bad butterfly-watcher, not often moving beyond the basics, though I remember a glorious moment when I was in Scotland and saw a butterfly I had never seen before. 'That must be a Scotch Argus,' I said, not quite facetiously. I looked it up – on an app on my phone – and blow me, a Scotch Argus is exactly what it was. I laughed out loud: life – wildlife – should be like that.

But there I was in Norfolk, and there was a dragonfly, but it wasn't a Norfolk hawker. Which was really rather a poor show.

Looking at dragonflies always seems to me a deeply useful thing: not least because it demolishes the Expertise Fallacy. Perhaps that's a disease of the technological age: the period that began when television first became a factor in human lives. You can switch on and see all kinds of experts: on politics, on economics, on art, even on wildlife. And while it's always good to learn, there's a rogue circuit that cuts into our minds: if you're not an expert there's no point in even looking for yourself. Delegate the task to the expert. I can't tell a song thrush from a blackbird, but David Attenborough and Bill Oddie and Chris Packham can, so I'll leave it to them. Why bother trying? Why bother trying to improve when I start a level so far below that of the experts?

We have forgotten that there is all the difference in the world between knowing nothing and knowing a little. The expert still knows far more than me, but so what? I have learned a little and that's not only a small adventure, it's brought me a little closer, not to the expert but to the subject that caught my interest.

My shaky knowledge of dragonflies is both an embarrassment and a thrill. Ignorance is an adventure – or rather, it's like Bag End, the residence of Bilbo Baggins: the ideal place from which to embark on an adventure. By trying and usually failing to identify dragonflies, I am spending a great deal more time looking at them, being taken up by them.

The dragonfly
can't quite land
on that blade of grass

A haiku from Basho, the 18th-century Japanese poet: one of his eternal vignettes. I suspect Basho was also a bad dragonfly-watcher.

Red pepper
put wings on it
red dragonfly

But this was not red but blue, with a pure, unapologetic blue tail, and when it landed – on that bladed iris leaf – it folded its wings up together, rather than holding them in two parallels, which meant it was not a dragonfly but a damselfly: in fact, it was a blue-tailed damselfly.

Life should be like that.

There was another and larger insect: a proper dragonfly this time. After a while I gave up trying to hallucinate green eyes and looked at the dragonfly for what it was: rather burly, amber in main body colour, and with a dark spot on each transparent wing. Four wings, so four spots. Hang on . . . isn't that a . . . four-spotted chaser.

Huzzah!

Wild June was off and running.

Home. Butterflies dance to the music of the sun.

A four-spotted chaser is different to a broad-bodied chaser, and a Norfolk hawker is different to a common hawker and a migrant hawker, and a blue-tailed damselfly is different both to a banded demoiselle and to a beautiful demoiselle (and that's not a gratuitous valuation from me, that's the name of the species: beautiful demoiselle). That's the point: and it's also the meaning of life. Life works not by making that same thing over and over again, but by making all kinds of different things. There'd be no point in a peregrine trying to hunt like Dusky and Pale, or for a kestrel to hunt at night. Better leave that to the owls.

Biodiversity. The central principle in life: nothing less. We appreciate this at a relatively deep level, and respond in our different ways: by revelling in the difference between a red admiral and a peacock butterfly; by trying to see as many species of birds as possible; by looking at mammals and wondering how it is that straw-coloured fruit bats and blue whales and northern sportive lemurs and elephants and lions and us humans could all be part of the same group of getting on for 5,000 different species.

> Is that a bird of paradise in the garden? Or a jay? Yes!

Here's one of Eddie's Wild June blogs. As usual, he dictated it to Cindy, who's always been his Boswell.

'I went on the marsh with Dad. It was warm and sunny, but the marsh was damp from the rain. Before we sat down we looked under the tins to see if anything was there, like we always do. And this time there was a grass snake. It was rolled up. We could see his head on top of all his coils. He looked at us for a minute then he slid away under the grass. He was extraordinary. He looked black – Dad said dark green – and he had a little yellow collar around his neck. I felt excited to see him. We enjoyed our sit. I had baked beans in a jar and pasta with cheese and tomato. We had a nice long sit in the sun and listened to the birds singing.'

The snake was a good three feet long. That's about the size at which a snake seems to me to become a proper snake. The movements – the lateral undulations – are bigger and slower and more deliberate, and it's immediately clear that these are uncompromisingly backboned animals, for there's nothing wormy about them whatsoever. For all their difference of shape, they're quite obviously one of us.

That sit was memorable for a bird that flew over. It was a tufted duck: not, you will agree, a bird that normally occasions much excitement. They are diving ducks, generally seen further out from the shore than mallards because they tend to exploit deeper water, looking more for animal life rather than plants to feed on. And there it was, a male strongly black and white, flying over in that whirring-winged, direct flight that ducks go in for.

I looked it up on the Marsh List and sure enough, it was the first tufted duck I had recorded from our place.

New bird.

Huzzah!

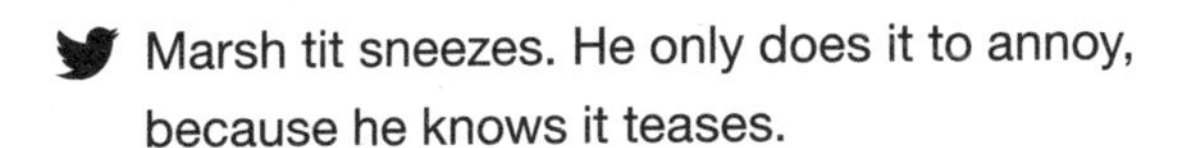

Early one morning the sun was shining; I was lain in bed, neither asleep nor awake, aware and unaware of the June plenty going on just the other side of the wall. The cuckoo was singing out still, hard at it. I had an urge to get up and join him out on the marsh; I had an urge to do nothing of the kind and the second urge won. At least, it won the gross physical battle. The spiritual battle was another thing, for my spirit slipped out of the window and went roaming round the marsh even as I slumbered and snoozed and thought about the day ahead and the chores I needed to do while relishing the pleasure of not doing them quite yet. Quiet savour of guilt: what a hog I was!

The cuckoo called again and again, that being the cuckoos' way. And I dozed again, that being my way, at least that morning. Come: you've got nature to look at, words to write and horses to attend to. Well, five more minutes, then. Cuckoo! No more than ten, all right?

Did I not hear a sudden rich bubbling? Certainly I woke from the latest doze convinced of it. Was that really a female cuckoo? And was it the same individual cuckoo that had been singing round the marsh for week after week? Almost certainly yes, for my money: but had he struck lucky for the second time?

I was fully awake now, and there was no convincing myself that the horses would wait contentedly any longer for their feed. And yet again, there was a story with the last page torn out. I always quite liked the What Happened Next feature on the television programme *A Question of Sport*: was it a goal? Or did he kick the ball clean out of the ground? Or was he rugby-tackled by a mad fan?

The cuckoo mystery was just like that, but Sue Barker would never show me the right answer. There's a passage in *A Dance to the Music of Time* in which Anthony Powell – or his narrator, Nick Jenkins – muses on the potential biographies of those who die young, stories which 'possess the mystic dignity of a headless statue, the poetry of enigmatic passages in an unfinished or mutilated manuscript, unburdened with contrived or banal ending'.

If you watch the same wild place and the shifting cast of creatures that live in it, you constantly come up against the same kind of mystery: all the more vivid for being forever uncompleted.

> Hooked, stogged, dripping in a tree like an abandoned umbrella ... marsh harrier in the rain.

There's often charm in our moral stories of non-human animals – the tortoise's victory over the hare, the robin who plucked the thorns from Christ's wounded head and who bears on his breast the colour of Christ's blood to this day – but they are stories about humans, not animals. (It's the same with Paddington, of course.) Modern notions of saintly dolphins and angelic choirs of whales follow a similar pattern. But it's not the job of non-human animals to tell us humans how to live.

It's not as if there weren't bad things happening out there in the loveliness of the cuckoo-echoing June. There were still remarkably few swifts. I looked at the sky and once again, I fretted. Of course, they might just be dodging the dodgy weather with a side-trip to the Bay of Biscay ...

If you learn how to look at a landscape correctly, you can see at a glance how much is no longer there. Almost, it's as if by loving wildlife you are wilfully bringing sadness into your life. My mother used to say that acquiring a pet was an investment in sadness. That didn't stop her acquiring dogs and loving them. And besides, what she said about pets is true on a much wider field: if you love anything that lives, you will have sadness in your life. That's the deal. Most of us accept it. We love, knowing that love will bring sadness.

That's because we also know that living without loving is not life.

But that shouldn't really be true of the wild world, should it? If we form an attachment to wild individuals, sure, they're likely to predecease us – but an attachment to the wild world itself really should be safe enough. The wild world itself is forever self-replenishing.

But it isn't, of course. To love the wild is to accept sadness as an inevitable part of your own existence. That's because we keep losing stuff: not as part of the eternal round of birth and copulation and death, but to the lumbering juggernaut of destruction.

I looked at the sky for swifts, I looked in vain, and I felt the pain of sadness. And, well, if that's the price you have to pay for loving wildlife, I embrace it willingly.

Above me a brief dagger and a flash of pink. It's like being a fish . . . kingfisher.

Richard the farrier took off one of Mia's back shoes. It bent almost in an S-shape as he did so. In the middle it was worn almost to the thinness of a knife-blade. Well, between us we'd been giving those shoes a lot of hard wear. All good.

There's a lot of hanging about, the day Richard comes. Mostly it's about being there in case there's a problem, and to keep the horses calm and content. So we talked the eternal horsey talk, never a hardship, and there were also periods of companionable silence, broken by the sharp rhythm of the hammer on shoe, on nail, on anvil. Every so often I go to and from the house. Richard likes his coffee more or less intravenous.

It follows that there is plenty of opportunity for looking at the sky above the marsh, so I keep a pair of binoculars to hand in case there's something worth a second look. The white stork, probably the biggest rarity that's turned up in our time here, flew over while the horses were being shod: there it was, powering on strongly from A to B, its neck and great red beak stuck out before it, long red legs trailing behind.

This time there was a red kite: a sudden instant vision of beauty: vivid sunlight making the bird glow red almost as bright as a robin, one of those little gasp-making moments. Basho would have caught this bird, this moment to perfection. I watched it, passed the binoculars to Richard, who enjoyed a brief inspection before picking up the rhythm of the hammer once again.

The red kites, as we have already seen, are another success story. So many birds of prey doing so well: hen harrier, red kite, peregrine, back from the brink.

The only blip in the pattern is hen harriers.

Cape Irago
nothing can match
the hawk's cry

Hawk? I bet that was a kite. Probably a black kite, rather than a red, but they are both very vocal birds. So yes: a vista of beauty, flying in direct from the 18th-century thanks to Basho.

But it doesn't have to be beautiful. And it doesn't have to be beautiful in order to be wonderful.

I have written a great deal about sport, and have often

argued about whether or not sport is entertainment. Well, sport can be very entertaining: compelling, dramatic, full of revelations of character and beauty. But for the most part the athletes aren't trying to create beauty, they're seeking victory. If they try to please the spectator, they betray both sport and themselves. Their duty is to victory and to the pursuit of excellence. And if this process is sometimes – or even often – entertaining, it's because sport is *incidentally* entertaining.

It's the same with the wild world. It is sometimes – often – beautiful, but its beauty is incidental. The creatures aren't out there trying to please humans: they're trying to get through the day by Not Dying, and they're trying to pass on their genes and become ancestors. There are no marks for artistic impression in Darwinian competition.

We tend to veer from one extreme to another when it comes to this question. We say that a wild creature – and by extension all nature – is either beautiful or ugly, attractive or repulsive, an example of what we have risen above or an example of what we should aspire to. The brutal nest-parasitism of cuckoos shows how far we have risen above nature; the perfect society of dolphins shows us how far we are from true grace.

Both ideas are equally wrong. Eddie is fascinated by bees and wasps: and if I were a better entomologist I would be able to point out to him the species of hunting wasps that use the marsh to make a living. These wasps care tenderly for the children they will never meet – a joyous example of selfless maternal care. The mother buries an egg and then furnishes the hole with a collection of small paralysed creatures: insects and small spiders. She doesn't kill them because she doesn't want them to rot. So she keeps them as a living larder: and the larvae hatch and eat their way through their slumbering snacks as they make the precarious journey towards adulthood.

You can find a conceptual beauty in this, I suppose. Or not: that's your choice. But it's not about us. It's about them.

> Five swallows came barrelling out of the stables. You know, I could have sworn that only two went in.

So here is a parable about beauty.

I spent two or three days of Wild June in Morocco, chasing wild birds. We were mostly there for Eleonora's falcon, a small and elegant bird that breeds in colonies along the migration routes of smaller birds. They breed late in the year and feed their chicks on birds that had been attempting to fly south on their way to warmer places. Instead they get a warm welcome from the falcons. (So the Eleonora's falcons are beautiful, if you like. And by being so mean to the gallant migrants, they are morally reprehensible, if you like. But it's not about you and it's not about me.)

I was travelling with Rod Tether, an old friend from Zambia. He worked for 12 years in North Luangwa National Park. Some of the most beautiful birds you could ever see are a daily pleasure in the Luangwa Valley: lilac-breasted roller, all iridescent blues; the cherryade colour of carmine bee-eaters, in hundreds and sometimes thousands; the softly bugling crowds of crowned crane; in the wet season the red bishop shines out like a live coal and the male paradise whydah somehow manages to fly with a tail longer than himself. Extravagances beyond our imaginings are a matter of routine. People come to see the big mammals and they go home birders. This is a place to inspire minds and change lives, as it changed mine.

The food in the tourist camps – always amazingly good – is cooked in open-air kitchens beneath thatched shades. Here the cooks bake bread daily in a tin buried beneath the fire and pull vast sticky cakes from the same unlikely place. One day in camp the kitchen staff called Rod.

'Come and see – come and see! We have a bird in the kitchen – and it is the most beautiful bird we have ever seen!'

They were entranced, in awe, knocked sideways by loveliness.

Rod hurried to the kitchen. He had no trouble identifying the beautiful bird, even though it was the first time this species had been recorded in the North Luangwa National Park.

House sparrow.

Meadow brown. Meadowsweet.
Sweet meadow.

There was a painted lady on the buddleia. And it was beautiful. Of course. Just the one, but welcome for all that. This is a butterfly you can never quite rely on, so it's always a small thrill when you meet one. I suppose they too are a parable.

They are boom-and-busters: sometimes – like 2009 – present in huge numbers; at other years, like this one, there were far fewer. They have one of those deeply complicated lifestyles: you learn the facts and wonder how the hell did they ever come up with that? The reason they did is beyond our understanding: it's Time. It's about Time deeper than we humans can understand it: Time not as three score year and ten, or Time measured in generations back to our grandparents' childhood, or Time back to 1066 or the arrival of the Romans. That is Shallow Time, historical time, Time as we humans understand it. It's not Time as life on our planet operates.

That is to say, it is Deep Time. The sort of time in which a millennium is an eye-blink. Over the course of unnumbered centuries of Deep Time, painted ladies evolved a lifestyle of constant movement. They migrate, but not as individuals. They migrate as a community, and they do so in an endless series of generations, forwards and back, forwards and back. It's been speculated that if we humans ever fly to the stars, we will do so in the form of a breeding colony. That's what the painted ladies do, travelling not from Earth to Proxima Centauri, but from Africa to Europe. They have reached Orkney, though not every year, and they have been recorded in the Arctic Circle, so they are intrepid beasts.

It used to be thought that the butterflies that bred in Britain were a doomed generation: a waste, an act of inevitable folly from the blind forces of evolution. Then it was discovered that the butterflies make a reverse migration: travelling back south again, and breeding as they go.

So that's another parable. Don't underestimate wildlife. Or evolution.

Outdoor supper is good even though the little owls heckle the conversation.

'Simon Barnes?'

'Yes, Mrs Holland.'

A twice-daily ceremony known as Taking Register. It happened first thing every morning and afternoon at Sunnyhill School.

Sometimes sitting out on the marsh is like that. The Maytime frenzies were over, but the breeding birds still sang out every now and then, just in case. Some species were thinking about and even already getting on with a second brood, and that requires a reconfirmation of self and of territory. The singing was now more intermittent: less like a mad chorus, and more like individual soloists taking turns, like everyone doing a party piece.

I didn't call out the individual names, as Mrs Holland did. It just seemed as if the birds were taking turns to announce themselves, so that I could make a mark in my notebook, acknowledge their presence and pass on.

'Sedge warbler?'

'Yes, Mr Barnes.'

The unbelievable unending complexity.

'Whitethroat?'

'Yes, Mr Barnes.'

A dry, hurried song from the brambles, surprisingly musical once you have come to terms with its nature.

'Chiffchaff?'

'Yes, Mr Barnes!'

Saying its own name tirelessly, again and again.

'Cetti's warbler?'

'Si, Signor Barnes!'

That great, jubilant shout.

'Willow warbler?'

'Yes, Mr Barnes!'

And that sweet, lisping descent down the scale, touching

my heart as no other song quite does. It's beautiful – at least I think so, for what that matters, for what I matter.

Six species of warbler. All present and correct, and presently and correctly breeding many splendid chicks who would, I very much hoped, return to the marsh and warble and breed in their turn. And the marsh was making this possible.

In June the year turns once again, tipping from hope towards achievement.

And how should this cease?

15

LIFE IS NOT TIDY

The air is filled with cries of falcons. The kestrels have fledged then.

At 05.24 BST the sun was directly above the tropic of Cancer. It was the June solstice. In London the sun was above the horizon for 16 hours and 41 minutes, a few minutes longer than that over the marsh. Midsummer: the longest day, the shortest night: perhaps the year's great moment of achievement.

I spent a fair bit of that day unconscious. I was at last having the surgery on my knee. When I came to I felt somewhat suboptimal, as if my bed was floating on a turbulent patch of sea. From my window in the hospital I could see trees. I remembered a piece of research about hospital windows: apparently you recover more quickly from a major operation if you have a window, and faster again if you can see trees. I could see the tops of a small stand of chestnuts: call that a good omen. I wouldn't be here long enough for the trees to make much difference: I'd be home as soon as I could stand, though that felt like a pretty crazy ambition at first. But still, the trees were vaguely cheering.

The knee had been pretty tedious, but now I had had the op, it was of course a great deal worse: the paradox of intervention. After a while I discovered that I could walk. Just. Damn it: I should have leapt off the operating table and danced like Blake's Glad Day.

But that was secondary to the effect of the anaesthetic. It seemed I was now living inside a cloud of shit. The instruction was no booze for 48 hours. Still worse was the fact that a drink was the last thing I wanted. Once I was home, with immense kindness, Cindy made me a sitting place on the veranda, where I could put my leg up and see trees. Eddie brought me my binoculars; he has an acute radar that seeks out the misfortunes of others, and he believes it's his job to try and ease them. He was the official assistant to the playground first-aider at his junior school; the children voted him as the recipient for their annual Peace Prize. He's always the first to hug a person at a time of bereavement, and when Thomas, who's blind, pays us a visit, Eddie is always there to offer an arm. So when I was briefly struck down, Eddie was there to help.

I sat there looking out at the garden and thinking about my grandfather. He was a marvellous gardener. In his garden in King's Heath, Birmingham a rockery cascaded down to a generous lawn that led to an extravagant rose garden. A kitchen garden lay beyond that, along with three beehives. Both sides of the garden were lined with blackberry bushes: a cultivated variety that gave thick, sharp, fat fruit. You were never short of honey and preserved blackberries in Vicarage Road. In my memory that garden covers many, many acres. The lawn was, of course, perfectly striped. It pleased my grandfather to look out from the French windows and savour the neatness: everything in its place.

I can't say that he wouldn't have approved of our lawn because he would. He loved new ideas and radical solutions: the idea of gardening for wildlife would have enthralled him. He might even have taken it up himself. All the same, there was an impressive contrast with the bowling-green lawn he cultivated and the shaggy expanse of long grass and wildflowers that lay between the house and the first line of dykes. It would have been nice to explain it all to him, though I have never threatened to be the practical gardener he was.

The way to make a wild garden look acceptable to human eyes is to mow paths. We like to feel that the landscape of home is under our control. We want it tamed – humanised – at least to extent. If you mow a series of paths across and around your areas of shagginess, it becomes clear to any observer that you have done this on purpose. You actually want it to look like this. It's not that you're lazy: this land reflects your own deliberate choice.

I looked out from my place of rest, right leg propped high, a cup of tea to hand, and I saw a bird fly low over the tall grass and take a perch in the old oak that stands on the right of garden, hard by the house.

I raised the binoculars, and at once I had a perfect view of a little owl, staring straight back at me, its yellow eyes glinting like a pair of little suns in his little cross face as he perched there in the first hints of dusk. He was clearly on a much-used hunting perch, and the life-filled long grass beneath him was part of his private hunting estate. Little owls will hunt in daylight when it suits them. I knew that this one had other beaks than his own to fill: there was a nest in the one of the big willows along the fenceline.

It was a gorgeous, glorious and inspiring moment: the bird so close, thanks to my stillness, and his eyes so bright, giving me a sort of death-ray stare that seemed to burn through the binoculars, gathering intensity as it travelled through all the lenses and prisms. You can damage your eyes if you stare at the sun through binoculars: I fancied that little owls could pull off the same trick.

There are no statistics on the way little owls expedite recovery from minor operations. But from my examination of a sample of one, I can report that the patient responded with good vibes and positive thoughts.

My 48 hours would be tomorrow. Maybe the smallest of small drinks would be an option.

> Are they red admirals on the buddleia? Or red air chief marshals?

In Ken Kesey's *One Flew Over the Cuckoo's Nest*, the narrator – 'Chief' Bromden, sometimes called Chief Broom – talks of the power of the fog machine, which the people in control of the psychiatric ward switch on to incapacitate the patients: presumably a reference to drugs used to treat the patients or to keep them quiet. I had been affected by the fog machine for some days after my op, but it was clearing. I was even limping a little less.

And of course, back to work: *scribo ergo sum*, and all that. Also *scribo ergo I can pay for the groceries*. Back to work as soon as I could get back down the garden to my hut.

How many notebooks on my desk? Two related to this book; three if you count last year's diary; four if you count Eddie's Wild June diary. Another one to do with some notes about magic I've been making with no very constructive aim in mind. Another with ideas for a novel I may or may not get on with: it might be nice to do another. Another notebook about wildflowers, which I started in an attempt to lower my levels of ignorance. Yet another containing a record of commissions and the names of editorial staff in various different publications. Two notebooks, both of which I use for, well, making notes: a big one that stays on the desk and a small one for travelling. A number of books not actually on bookshelves: about 40 altogether, at a quick count, five of which I am reading for interest, three of which I am reading in order to write a review, others I have consulted and not yet returned to the shelves, still others that are lying about for no good reason at all that I can remember. Some fossils, two of which I found myself. A bowl Joseph made many years ago, now full of paper clips. Receipts. Blank paper. A set of proofs from another book. Binoculars. Bat-detector (what's that doing here?). *Dad's Army* mug (Don't Panic!) full of pens.

You get the idea. Untidy. But it's an untidiness that reflects busyness. You can say what you like about my organisational skills, but if you step into this hut, you know that this desk and its surroundings are used. This is a working place, and it's a space that's full of life.

It's the same with the garden and, to an extent, the same with the marsh. Untidy, sure, but full of life. If I had mown the grass down to bowling-green length and chemicalised it to monocultural perfection, there would be no small mammals and large insects in the grass – so there would be no little owl in the tree. The owl was a tribute to our purposed untidiness: to the way in which we have purposefully loosened a little control.

Untidiness is life. Tidiness is death.

You visit a friend's house. It's tidy, of course – why isn't your house always tidy like that? – but they finished tidying it ten minutes ago, knowing you were on your way. A living house is a tribute to the life that goes on inside it: therefore it's seldom seriously tidy. Toys, garments, one of Joseph's guitar-cases, a palette and paintbrushes in the sink, newspapers, books. Did you mean to leave the flour out? Yes, I'm making pizzas. Shouldn't those books be down in your hut? No, they should be back on the bookshelves up here, but I can't get to the shelves because of all the stuff in the way. Eddie, what have you done with your shoes? I do wish you'd put things away sometimes. And preferably in the right place . . .

That is the nature of our house, of most houses. And it is also the template for a living countryside. Not hedges flailed to buggery or grubbed out altogether, not trees mercilessly pruned or cut down, not fields ploughed right to the very edge, not gardens concreted over for car-space, not green spaces with a furiously mown sward and trees like lollipops.

These are all tidy. These are all very ordered spaces. And to a greater or lesser extent they're dead.

There's order out there on the marsh and there's order in the wildness of the garden. But it's not human order. It doesn't look like human order or feel like human order. The order of the marsh is deeper and more serious than anything humans can dream up or create with machines and chemicals. It's life. And life is a mess.

Then a sudden din: three oystercatchers charging across the sky, piping and squeaking at top of their voices, and for no good reason apparent on the ground. Because they can, no doubt. Are these birds new-fledged? That'd be my bet. They flew across, making the place took untidy.

There's an old euphemism for untidy: lived-in.

On a sun-basking morning the path is paved with butterflies.

Lying in a hammock is a deeply stressful experience. That's because it's accompanied by an eternal and unsolvable dilemma. Do I place my hat over my face? Or not?

Hammocks should be sold by specialist wildlife suppliers. The Natural History Book Shop has sold me books, the bat-detector and refugia – tins – for reptiles, and they offer many more excitements that bring you closer to wildlife. But they don't sell hammocks: and the hammock is as good a device for studying wildlife as the moth-trap and the entomologist's pooter.

In the warmer months the hammock lies in the shade of the big ash, a few limping paces from my hut. It hangs from its own frame and offers the sublime comfort that only a hammock can bring. But it also brings this appalling dilemma. The strain of this piece of decision-making almost overwhelms me every time, but in the end, I rise to the occasion like a man.

It's a delicate calculation based on the brightness of the light and the mood of the hammock-born human. So I was lying there hatless, which means you can open your eyes and look at the sky, and that's always worth doing. There's a lot of sky to choose from round our place, and you never know what might turn up in it.

This time there was the classic birding moment: long-winged, elegant, silhouette like a huge swift. That, as birders will at once recognise, is the standard – and vivid – description of a hobby in flight: a falcon with wings that are long, sharp and slender even by the standards set by fellow falcons. They come to Britain for the warmer months to breed, and they are virtuoso fliers, specialising in the most difficult kinds of aerial prey.

They are adept at taking dragonflies from the air, and they eat them on the wing; occasionally, especially in late spring, you will see a party of them over a wetland, circling and circling and helping themselves to flying dragons. But they will also take swallows, martins and swifts, which always seems to be going out of the way to make trouble for themselves. I know! Let's find the fastest and most agile of all flying birds – and then specialise in taking them on the wing.

They will make the stoop, in the falcon's anchor shape. They're not as fast as peregrines, but they are highly manoeuvrable: well, they'd have to be, wouldn't they? They are a slightly more esoteric version of the swifts, in that they're globe's-still-working birds. Seeing a hobby always seems like a good omen.

One of the plusses of the hat option is sensory deprivation: if you're not distracted by flying falcons, you're more likely to get a doze. But that theory doesn't work in a decently wild spot because when you deprive yourself of sight you emphasise sound.

And so, with the sun squinting through the latticework of my hat, I lay suspended between heaven and earth, and the most glorious music came bursting out: a solo of low-key extended virtuosity. At first I thought it was a blackcap, but as I got my hearing into focus and my brain more fully engaged, I realised it was nothing of the kind. It was a garden warbler: no doubt the same bird that was singing earlier in the year, still present. There was a good bet that he was half of a breeding pair: which would make seven species of breeding warbler in these scant few acres.

A new one for the register, then.

I am the master of horizontal birdwatching.

> I leave my writing hut and the first thing I find is a comma.

There is a sort of extended lean-to around the kitchen door, copiously windowed, with one closed-off sitting place and then another sitting place out in the open. Both can work as inadvertent traps; freeing confined animals is an accepted part of family routine. Birds get stuck in the closed part. Usually it's enough to open all the doors and wait for them to find their own way out. If not, a little gentle guidance will normally get them out. Just occasionally it's necessary to catch one. It's a tremendously difficult thing, holding a small living bird in your hands: you are acutely aware that too much pressure will crush the poor thing, but if you don't apply enough, you will lose the bird, making it more frightened than ever, and twice as hard to catch. Once a cock pheasant got stuck inside: I caught him in the manner of a scrum-half and made a dashing pass back into the wilderness; the creature thanked me by drawing blood from my left thumb.

The more open area is no trouble for birds, but big insects get trapped on the translucent ceiling, caught in the angle of beam and plastic roof. Eddie, being a caring person, is particularly vigilant here: as soon as a dragonfly gets caught up in the architecture, those big, brittle wings fizzing and drumming on the roof above, Eddie will summon one of us to show the way out. There's always a net leaning against the wall on its long bamboo handle, but freeing dragonflies is not always an easy task. They are highly mobile, with excellent vision and they're very wary of the net. Often you will move them two, three beams closer to the open, and they will perversely fly back in again. Eventually you succeed: and the dragonfly is off, trying to make up for the time lost beneath our roof. Keep a watch for that hobby, fella.

'Dad! Come here!'

'Why?'

'Big insect!'

'Dragonfly?'

'Don't know.'

Meaning it probably wasn't a dragonfly. If it was he'd have known.

I'd been sitting out in the garden, and I was too comfortable to get myself any refreshment. I got up with slight reluctance – and found the insect of the year.

It was up there among the beams, baffled by the fact that you can see through the roof but you can't fly through it. It seemed an impossible thing at first: a small furry rugby ball – a small furry *striped* rugby ball – suspended in the air with no immediately visible means of support. After a slight adjustment of the mind I could see the wings as a form of mist. There was pink on the fur of the body and in the wings.

I got the net, but I wanted to spend a few moments looking at the marvel before I performed the brief miracle of freedom.

'It's a moth, Eddie. Remember the moth we saw on the flowers when we were on Alderney?'

A huge effort of memory.

'Humm . . . humming . . . hummingbird!'

'Brilliant, Eddie! You're such a good observer. Hummingbird hawk-moth!'

Every summer the RSPB and other wildlife organisations get calls from people thrilled to report a hummingbird in their garden. In a way, the creature they have just seen is a greater miracle than a hummingbird crossing the Atlantic to sup nectar in the wrong hemisphere.

Hummingbird hawk-moths have adopted the same strategy as hummingbirds. They have highly developed hovering skills as well as the ability to fly backwards. They exploit these talents to feed from flowers, without resorting to the banal expedient of perching. This freedom from the need to find a foothold makes many more flowers available to them: they can feed on nectar despite their bulk. If either hummingbird or hummingbird hawk-moth tried to alight on their food sources, their weight would often be too much for the plant. So they both feed on the wing. In other words, the same solution to the same problem has been reached by a radically different evolutionary route: like swifts and swallows but still more dramatic, since the relationship between the two is incomparably more distant.

It's called a convergence, or convergent evolution: insects, birds, pterosaurs and bats all – quite separately – evolved the ability to fly, and they all use quite different bodily structures to do so. Perhaps still more intriguing, squids and octopuses evolved intelligence from a quite different route to us vertebrates.

Eddie called Joseph and Cindy for a moment to enjoy the moth, and then I took the net and eased it from beam to beam – their tendency to fly on the same spot made this comparatively easy, much easier than those hot-rod dragonflies – and soon it was out in the open. I was hoping it would start feeding from the buddleia or the other nectar-rich insect-tempting plants we have in the garden, but this one was clearly glad to get away. He switched from hover to straight and level flight and went hammering off towards the marsh.

Hummingbird hawk-moths turn up in this country in most summers, more frequently down in the warmer south, crossing over from the continent. This was the first we had seen at our place: and it made a resounding conclusion to Wild June.

Is that a squadron of flying cats or have the buzzards just fledged?

One of the sudden pleasures of this season is the fledge-out: the day when quite suddenly the place is full of birds, all in a gang together, all the same kind, all very eager to get on with life but not quite sure what birds are supposed to do. 'Someone's opened a can of blue tits,' Cindy remarked. It was as if blue tits were about to conquer the world, reach plague proportions, drive humanity out of house and home. It's an illusion, if a merry one: it's all about numbers, and the fact that few if any will survive their first winter.

Not that you think such grim thoughts, of course: and anyway, it's not really all that grim, or no grimmer than most facts about life. If one blue tit, in the course of its brief lifetime, succeeds in fathering or mothering a chick that goes on to raise chicks of its own, then it has won life's lottery. Many species adopt a strategy based on large numbers: codfish lay eggs a million at a time with the same ultimate aim in view. Tsetse flies, as I routinely explain to clients on safari in Zambia, are not evil monsters. They are a touching example of maternal care: the female will raise a single grub inside her own body. The grub will go through three larval stages, nourished by a milky substance that the female secretes. It's about the trade-off between parental care and numbers, and that's about assessing the odds and responding accordingly. Any bookmaker would understand evolution in an instant of time.

Then the swallows in Norah's stable fledged, and if it was not the banishment of worry about that species, it was certainly an un-turn-downable invitation to set worry aside for a while. There over the meadow the swallows were making their first attempts at being intrepid intercontinental aeronauts: working on the circles and esses and arabesques and spirals that are the essential life skills of the swallow. I remember Joseph – never a great one for crawling – ambitiously taking his first steps while Cindy and I surrounded him like slip-fielders.

The swallows were making their baby-steps: tumbling from the nests and in an instant becoming accomplished whizzers and whirlers, as if Joseph had taken his first steps and instantly danced the lead in *Swan Lake*.

I once watched a fledge-out when the young swallows decided – or were shown – how swallows drink: swooping low over the water and taking a sip, wings raised high to keep them dry and to facilitate the brief shallow glide, and as they reached the water's surface they ducked their heads and drank. It was gloriously comic. Sometimes the young swallows missed altogether, timidly going in too high, sometimes one would go too deep and crash: tumbling and splashing before fluttering away in a shower of sunlit drops. It was a fine thing to watch: showing that the flying skills of a swallow are only innate to a certain degree; a good deal of it also has to be learned. To what extent were the parent birds deliberately showing them what to do? If at all? Either way it takes hard work to be a swallow.

And then a fledge-out of kestrels. I've never been able to work out where they nest: some way off, I suspect. But there were three of them flying over the marsh: not really a serious bit of hunting practice, rather a joyful celebration of kestrelness: why look for food when you've got a sibling you can swoop on from a dizzy height? They crossed the big sky, playing Spitfires and Messerschmitts, exulting in the new life flowing their veins.

This detonation of new life felt for moments at a time like

being young again myself: full of bounce and vigour and no idea at all how to organise such things, only the vaguest inkling what human beings actually do but eager to get on with it – well, parts of it – as soon as possible. Friends! Beer! Books! Travel! Adventure! Girls!

Perhaps all those fledged-out swallows went back from loop-the-looping the meadow to receive the congratulations of their parents. Well done, little one! And in a few weeks, we'll have another little flight. Cape Town!

In the vastness of the Norfolk sky even a crane is small and hard to find.*

It's rather throwing to share your name with a friend. Simon and I also shared a flat sometime back in those days of being young. It's always good to see him, if only to see what mad craze currently possesses him. Simon might be classified as a satisfied Toad. When Toad took on a new craze, he invariably made a hash of it: but when Simon takes on something new, he generally masters it. Few people have worked as a top-end chef, a potter and a website designer and done all three triumphantly.

He arrived and immediately gave us a handful of wooden spoons, all of which he had carved himself with delicate strokes of a curved knife. He told us he had brought a drone with him, so the next morning – after a long evening of talk aided by a little drink – he launched the drone over the marsh and surrounding countryside; his work provides the basis of the map in this book. The pictures also went to

* A reference to *Big Blue Whale* by Nicola Davies.

the parish because they own the land next door where Jane grazes her sheep, and to the Norfolk Wildlife Trust. They were good pictures, and the drone was handled with notable competence: what you'd expect from that Simon, though not from this one.

He and I spent an afternoon on and around Hickling Broad, hiring an electric day boat. This is lovely way to travel: almost completely silent, so the reed warblers, sedge warblers, reed buntings and bearded tits could be heard with perfect clarity. And as we travelled, sometimes talking, sometimes enjoying the silences, a pair of cranes flew across the reeds before vanishing.

There is a real thrill in their unEnglishness: flying with long neck stretched out in front of them and longer legs trailing behind, and five feet or more from end to end, they seem impossibly exotic. If any bird symbolised hope in the wild wetness of Norfolk, it's the crane, back with us after 500 years of absence. If we can undo half a millennium of harm, we have scope to put a few more things right.

Would cranes be the ultimate wonder for our own stretch of marsh? Perhaps so. Cranes as regular visitors ... cranes as breeding birds ... well, no one can say it's not possible. It hardly seemed possible that cranes would return to England after five centuries of absence.

I looked out at our own patch of marsh: particularly at the lower, drier, less scrubby part that we bought from Barry almost a year back. It didn't take much hard work to see a pair of cranes high-stepping across, dipping their long necks to peck in their fastidious way at the life of the place. You never know what wild thing will turn up next in a place – so long as the place itself is suitably wild.

Outdoor supper is good even though little owls heckle the conversation.

There are times when I think the greatest blow to civilisation in the 21st century is prosecco. It's a nice drink, but it's not a substitute for champagne.

There are certain kinds of birds – cranes, for example, or Savi's warblers – that are undeniably Special. The idea of specialness is very dear to us humans; it's part of the human condition. We took the 28-day lunar month, divided it into four parts of seven days each, and in order to come to terms with the rhythm of passing time, we made one day Special. When I was a boy, Sunday was church followed by roast dinner. These days on Sunday evenings we sit down to a great curry feast: I cook it myself and love the lingering smell of spice across the house.

Champagne is used to mark special events: birthdays, wedding anniversary, a new bird for the list, the publication of a book, the sale of a significant piece of art, and so forth. But the great thing about champagne is that it also works reflexively. It can be used to mark a special occasion: but it can also be used to make any occasion special.

And so it was that evening. Cindy and I sat in the garden, the light fading without hurry. Pop and clink and sip: golden wine on a golden evening. This was, as I remember, the cheapest champagne sold by the Co-op. The temperature was perfect – of the wine, of the evening: a rare time of stillness. It was as if the marsh, as if the year itself had taken a moment to pause and appreciate how far it had come: how long it was since the chill days, when it seemed that the world must be forever cold.

And then the yelping began: a hooting, rather exuberant kind of yelping, and with them came the little owls. If this wasn't the actual moment of a fledge-out it was pretty close: the little little owls were still fluffy at the edges with their baby down. They looked delightfully absurd as they lined up along the fence and tried to do the things that little owls are supposed to do.

There are moments of perfection: complete and unabashed. They come rarely and fleetingly, but when they do they often have that weird stretchy quality that destroys time as an objective quantity and makes it a purely personal affair. Such moments come more readily when they follow a glass of champagne drunk rather quickly. I poured again: this one to be drunk more slowly. Us, the marsh, the owls, the bubbles, the yelps, the silences: a moment lit up with the flames of eternity.

16

The Year Holds its Breath

> Cancel the morning walk. I need protection from the storm of butterflies.

The moment when the hidden becomes suddenly and dramatically visible is an experience known to us all. It's associated with moments of glory and moments of horror. It's something all film directors work on: the sudden revelation. And it's part of the routine of wildlifing: now you don't see it, now you do. I remember finding the carcass of a majestic male otter by the roadside: good news and bad news. The bad news is obvious; the good news is that the sad find revealed that otters were quite clearly back in the local river system.

So when Carl asked if I would like to join him on a visit to a dead whale, I was up for the treat. It had beached itself in North Norfolk, near the great RSPB reserve at Titchwell. Carl is Norfolk's cetacean – whale and dolphin – recorder, so it was his job to take measurements and a DNA sample and to gather other information.

It wasn't a massive creature, though big enough: a good 17 feet long, we ascertained. I had the important job of holding the other end of the tape measure, and it was a task better performed on the windward side. After a fair amount of thought, Carl was pretty convinced this was a Sowerby's beaked whale; the DNA test later proved him right. They are extraordinary creatures even by the standards of whales. They dive – and they dive deep. They can stay down for half an hour, holding their breath while they hunt for squid, which they consume by sucking them in with toothless jaws.

In the sad, sagging flesh it was possible to make out the ghost of the streamlined, athletic creature this once was. Carl thrust his gloved hand inside the whale's mouth and found no teeth. Males have two teeth, used exclusively for fighting, rather than squid-eating. So with the information recorded, we retreated back into the fastness of Titchwell for a few deep breaths and a few nice birds.

There have been 30 species of cetacean recorded off the shores of Britain, and 20 in Norfolk. So next time you look out to sea, think of the fabulous and enormous creatures that live there out of sight, yet breathing the air just as we do. We can look out at the sea and have no idea that they're there at all: no awareness of their existence, no understanding of what kind of damage we may be doing to them.

> Sky-writing lessons for young swallows. They're getting the hang of S and O.

Cindy found a shell alongside one of the dykes. I was thrown at first: it looked like the kind of seashell she sells on the seashore. That is to say, it wasn't like a snail shell: a univalve with a single door. It was a bivalve, like an oyster or a scallop, with two doors held together by a hinge. Was this something pushed in by the complex movements of tides and current? Or was it a souvenir of the long-retreated seabed? Then the penny dropped and the heel of the hand struck the centre of the forehead with a hollow thud: of course! I knew it all along. It was a freshwater mussel: we have six species in Britain.

The shell was a supremely graceful oval, a luscious line that Barbara Hepworth would have loved and probably used, perhaps 1,000 times larger: I could imagine it standing tall and proud on the banks of the Thames by the Oxo Tower. This shell was one of those impossibly lovely things that the wild world throws up on a routine basis.

I had no idea there were mussels in these surrounding waterways. Like most of us, I had never given it a thought. And yet they matter, these humble and lovely things. Not only do they indicate that the water in which they live is in good shape, but they also help to keep it that way. They are one of those keystone species: species whose presence or absence plays a crucial role in the entire ecosystem in which they have their being. Keystone species can include large carnivores, beavers and these bottom-dwelling beasts.

That's because they are siphon-feeders: a big one – some can get up to six inches across and live for 100 years – can process 40 litres of water in a day, so a big colony changes the nature of the waters they live in: taking out nutrients, algae, bacteria and pollutants. There can't be many on the river: it's routinely dredged to keep it open for boating. I wondered how many there were in those wide, generous dykes: had the agricultural chemicals and the passing river-traffic done for them? How old was the shell I held in my hand?

The lives of the adults are not too exciting: they mostly stay in a single place, siphoning away. But they have a wild youth, living as quite another kind of creature. There are species of mussel larvae that produce long sticky strings, and those that win the lottery manage to snag a fish. They then winch themselves up the string, attach themselves to the fish and operate as blood-feeding parasites till it's time for them to drop off and start life as siphoning adults: another classic example of the extraordinary things going on beyond our sight and below the threshold of human awareness.

Did this mussel shell mean that there was a living, siphoning, fully functioning population down there? Or was it an ancient thing that had turned up by chance, a souvenir of days long gone?

Tea with butterflies. No milk, no sugar. Just brimstone, please.

Here is another parable: the parable of the partridge.

I have an intermittent and fascinating – I'm the one being

fascinated – correspondence with Philip Howse, author of, among other works, *Seeing Butterflies*. He is fascinated by camouflage. He has taken the idea of the peacock butterfly's mimicry of an owl and radically extended it. He occasionally sends me an image of a butterfly or moth: 'To me this looks extraordinarily like a small mammal with a pink nose. Any thoughts?' The other day he sent me an image of a butterfly that was exactly like a tiger seen from above.

One of my earlier thoughts in this correspondence was 'just how much LSD did you consume in your wild youth?' The answer was none whatsoever: he is just extraordinarily gifted at seeing. And he supplied me with a parable about hidden life.

It concerns Hugh B. Cott, zoologist and expert on natural and military camouflage; he played a significant role in the Second World War. He listed nine different kinds of visual deception: I list them here because it would be a shame not to.

1. Merging, like hares and polar bears
2. Disruption, like a great-spotted woodpecker
3. Disguise, like a stick insect
4. Misdirection, like a peacock butterfly
5. Dazzle, like some grasshoppers
6. Decoy, like the lit-up lure of an angler fish
7. Smokescreen, like the ink of a cuttlefish
8. The dummy, like flies and ants
9. False display of strength, like a toad inflating itself

Cott spent a day trying to photograph the cryptic camouflage – an example of category one – of a female grey partridge on a nest. He had staked out the nest and set up his camera and waited all day for her to return to the empty nest. No show. Eventually, he gave it up as a bad job, but he took some photos of the nest anyway, in case it should be useful.

When he examined the resulting photographs he saw that the partridge had been on her nest all along.

> I think it's called a horsefly because it's nearly as big as my horse.

It was the moment when the year holds its breath. It's like the moment between thought and action, stretched out over a matter of weeks. It's the heady pause between high spring and the urgent movement of migration. The marsh was not exactly silenced – never that – but it was quiet: bursts and murmurs of song. The hard work was done.

Birders talk about the 'summer doldrums': the time of year when there's not much going on. The great peak of the Maytime breeding frenzies are long gone; the craziness of October, when anything might turn up anywhere, was a long way ahead. And so birders turn to butterflies and dragonflies while they wait for things to start hotting up again.

I have never found this lessening of intensity a difficult time myself. I am more interested in the birds' rhythm than the birders' rhythm, though I intend no snobbery in this. Crash-hot ID-people need to exercise and hone these skills, and a quiet season makes them restless.

What I like is the productivity behind the quietness. It may be a quiet time for you and me; for the birds, it's just about the most important time of the year: finishing the job of breeding, fledging the chicks, going further along the road towards becoming an ancestor. How many of those chicks will survive the winter? Some will spend it nearby, some will go on short journeys, some will cross half the planet. How many of those chicks will survive and breed in their turn? This is the moment of achievement, the year's biggest moment of all: but it also asks a million questions. Or maybe just one: what happens next?

And yet, even as this most urgent of all questions must be asked, there is an air of quiet repletion about the marsh and the world. Eddie took supper on the marsh and I accompanied him. It was an evening for contemplation rather than urgent looking and listening . . . but what were those ginger beetles crawling up the stem of an umbellifer? I knew them: soldier beetles, small and relatively long, rather square in shape. Perhaps they look like little guardsmen on sentry duty, and their red jackets are traditionally military. Not that they were acting in a particularly military fashion: they've acquired the nickname of hogweed bonking beetles, and they seem to spend their entire adult lives in lovely copulation bliss on bliss. In rare breaks from this activity they feed on pollen, aphids and nectar and are useful pollinators. And then—

'Eddie! Look here!'

He looked at it for a while, and then the cogs meshed gloriously: 'Hummingbird hawk-moth!'

That's one of the things about all comparatively unusual creatures: when you see one they act as if it is the most natural thing in the world. Nothing could have been less self-dramatising than the insect's meticulous, even fussy attention to the flowers of the marsh: a sip here and a sip there. I had once, ludicrously, taken part in a port-tasting (after a while they all tasted of cough medicine) in which the finest wines available to humanity were sipped, swilled round the mouth and then spat. The moth had the same seriousness of demeanour: flap-flap, sip-sip, and now let's try the next one, please.

It was a delight to see one untrapped, doing what hummingbird hawk-moths (and for that matter hummingbirds) do with such aplomb: such huge aeronautical skills used with such nonchalance. What are you staring at? It's what I do. It's what I'm *for*. But Eddie and I watched, entranced. There was a soft, almost reflective little hint of song from the willow warbler in the sallows.

Eddie and I decided to walk on and sit for a while on the second set of benches. It was that kind of evening. As we arrived at the pond – the pond that we decided was not to turn into a reedbed – there was a female mallard, but she made no immediate decision to spring into that splashing vertical take-off from water that mallards are so good at. The reason for this was clear a second or two later: ducklings!

Count them, then. Three, four – no, five! And then there was a sixth doing that sudden catch-up thing that ducks do, as if it had suddenly mounted an invisible underwater bicycle, and it pedalled like Chris Hoy storming the finish line in the velodrome, back into the group.

Few things are more enchanting than a hatch-out of ducklings: not fledged yet – of course not fledged by a long way – but fluffy little scraps of life, floating high on the surface like bath toys. The sight of creatures so tiny and so young – a few days at most – actually swimming is always rather disconcerting. Yes, of course you know intellectually that ducks are born to swim: but that they can leap straight from the egg into the water and do it with such confidence is boggling and beguiling. It's like the shock you feel when you travel to France and find that everyone – even very young children – can speak French. And there was a small example of management in action: the open water that we allowed was now allowing the mallards to live their watery lives. It was an infinitely pleasant savouring of the God illusion: as if we had created those ducks and had seen that they were good.

That night I was awoken by the sound of oystercatchers. A restlessness had infused them. They would be on the move soon, if they weren't already gone. I suspect that in that nocturnal call they were declaring the summer doldrums closed.

The rooks in the rookery murmur their togetherness. Esprit de corps.

Mid-July – my birthday still a few days off – and the oystercatchers had declared the beginning of autumn. Shocking how early this season of movement and change is upon us. When Carl and I had paid our respects to the late Sowerby's beaked whale, on our walk back through Titchwell reserve we had seen a delightful gathering of golden plovers. They were still in their summer breeding plumage – black below, dazzling spangles on their backs and wings – and yet these sumptuous colours were beginning to fade, were already a little rough round the edges.

They had gone over. They had completed their breeding: they were on the move. For them, autumn had arrived. Just think: all those people getting ready for what they thought were their summer holidays, unaware it was already autumn.

Jeremy Sorensen, former warden of the RSPB's Minsmere nature reserve in Suffolk, used to claim that summer didn't exist at all. Looking at the world in his entirely bird-centred – avicentric – way, he was convinced that there were three seasons only: you might call them breeding, dispersal and survival. We were already in the second of those two seasons, so far as some species were concerned. Those that bred were beginning to return to the life less hormonal, ceasing to sing, abandoning life as a half of a pair, often becoming part of a flock instead. An entirely different way of life, almost as radical a change as the metamorphosing freshwater mussels. For a few weeks the males are tough, amorous, protective, ever-ready to risk all: next they are self-effacing, meek, concerned only about themselves. Many species adopt quite different modes of existence for different seasons: some even look completely different. To all intents and purposes they really are different animals: and will remain so through the long months of survival. And if they do indeed survive, the males will change back into the rampaging, singing conquistadors, the great lovers and fighters they had once been.

We'd be off on our own autumn holidays soon enough. And Eddie was leaving school: ahead after the holidays lay three days a week in college, learning life skills, and another day at Clinks Care Farm.

The season of the summer doldrums is the season of change. Like all other seasons.

Here is another parable: the parable of the robin in the house.

There's always something slightly terrible – cosmically wrong – about a bird indoors. We once had a bloody great carrion crow in the hall: that was exciting. He was so bewildered – or was he? – that he let me wrap him up in a tea towel and pinion his wings. He remained perfectly still as I carried him outside. They're birds of considerable intelligence – every bit as smart as an ape, some scientists have claimed. So what was going through his mind as he allowed me to catch him, hold him firmly enough to stop him escaping, and then let him go, tossing him skywards with a great din of feathers as he took off without a caw of thanks? Perhaps he was shocked into apathy. Or perhaps he worked out that submission would lead to freedom. Who's to say?

The sitting room is mostly conservatory, glass-roofed and walled, so it was hard for the robin to find a way out. The last thing you want to do is panic the bird and have it crashing into the windows that it mistakes for moving air. Eddie was deeply alarmed: something about flying things very close to him makes him jump. The bird is trapped in your human world, and to stay is death.

All doors, all windows open, Cindy and I at opposite ends, trying to will the bird towards the double door, hoping that it wouldn't fly over our shoulders and head for the stairs. You want to explain the principles you're working on, so that the bird knows you have nothing but its own good at heart. The sound of fluttering and scrabbling, the occasional sick clonk as the bird strikes glass … and then it's as if the obstructions had dissolved, fallen away, turned to dust, and the bird is through the door, away from humanity, barrelling away at top speed, waiting to find a safe perch where it can pause for a moment and wonder what the bloody hell that was all about.

When we humans enter the wilder world it's a great and gratifying adventure. When a wild bird enters the human world it's pure horror. Even when the humans are on his side.

Birthday message from my new moth-trap: there are tigers at the bottom of our garden.

There's more than you think. That's the essential truth of nature. It's true even in these depleted times, and it's a truth that reflects the inadequacy of our thinking. Our minds jump to the assumption that if we are unaware of something – or unable to understand something – it doesn't matter. It doesn't even exist.

Cindy gave me a moth-trap for my birthday. I thought this was a nice idea, but the tiniest bit misguided. I knew very little about moths. I knew they existed but I wasn't aware of their existence in any meaningful way. I was by no means convinced that the moth-trap would be a major asset to personal, still less to family, life.

A moth-trap is basically a box and a light. The light is eerily bright and runs on mercury vapour. The box is built on the principle of the lobster pot: easily in, but not so easily out. Once inside, the trapped moth finds a lot of egg boxes, and creeps into one of the compartments to wait for release.

We talked it through with neighbours John and Louise and, being generous people, they said yes, of course, go ahead. Now they wouldn't think it was the light from an alien spacecraft at the bottom of the garden. We walked back to the house on the evening of my birthday with the glow of mercury uncannily shining out in the twilight. I wondered what creatures would be summoned by its spell. I also wondered if we would have any idea at all what we might be looking at. I had the right books; what I lacked was the right skill, the right knowledge. I didn't even know where to start, which page to start looking.

There are times when almost complete ignorance is thrilling: or rather, confronting that ignorance is thrilling. Stepping off a plane in a new country, that sort of thing. You realise how much more there is than you could possibly know. Dante is sometimes said to be the last person to possess all the knowledge and understanding of his culture: even as the renaissance was beginning, here was the very last renaissance man. These days all of us have vast areas of ignorance all around us. We can never be Dante: but trying and failing is a worthwhile exercise.

Dedicated moth-ers have told me that visiting the moth-trap of a morning is like coming downstairs on Christmas Day. I was wondering if it would feel as if I had been given a library of books about astrophysics: glorious but altogether beyond my scope.

Eddie was excited about the possibilities. On our trip to Alderney he had taken part in a moth morning. He is wary of creatures that fly too close to him, as the robin in the house showed. There was a time when he was alarmed by butterflies. It's understandable, when you think about it: one moment they seem to be part of a flower, next moment they are tearing off. It's that moment of transition that seems so unlikely, and to Eddie, so alarming. But he was so intrigued by the Alderney moths that he forced himself to look and admire. He liked the moths for themselves and also for the way they set off his own increased daring. He got close, but backed off when offered a moth to hold on his finger.

So we made our visit to my new moth-trap on the morning after my birthday. We were prepared for disappointment and damp squibs ... and got something close to a nuclear explosion.

The very first moth we saw was enormous and beautiful and its picture was actually on the cover of the field guide to moths: so we not only had wonder, we had wonder with a name. There, in gorgeous shades of pink and green, was an elephant hawk-moth.

Now, to proper moth-ers this is all like boasting that blue tits come to the nut-feeders, but to us it was as if a phoenix had come to the bird table. We could have stopped right there and it would have been a day of wonder: we could have thrown away the moth-trap at that moment and it would have been the most wonderful present.

But we didn't. We opened up the egg boxes and found more.

Hawk-moths are impressive insects: so big they are sometimes referred to as honorary birds. And there among the egg boxes were four more hawk-moths: creatures with uncannily rigged and scalloped wings. They looked like scale models of flying machines to be flown by magnificent men. These were poplar hawk-moths: and there was one perched on my finger almost as big as my hand.

And there was a moth on Eddie's fingers. After a series of advances and retreats, he went for it: and there sat this monster insect, with Eddie looking at it, awed by the moth, awed by his daring, and behind him stretched the marsh from which these fine creatures had all been recruited in the hours of darkness.

Who knew they were there? Not us, that's for sure. And there was the gaudy and lovely moth called a garden tiger: a moth with Bridget Riley op-art wings that, when folded forwards, revealed hindwings of black and orange.

We photographed most of them and released them all,

with care, into the surrounding bushes: the idea of moth-trapping is not to give a free meal to the birds – let 'em work for their living like everybody else. Some we identified quite easily, some we identified after a good old search, and a good few we had to give up on. It's about learning patterns, and that's not an overnight business.

Eddie wanted to take this further, and so with Cindy's help he drew a garden tiger and an elephant hawk-moth, colouring them himself with absorbed accuracy and neatness. And I was lost in wonder that there should be so many creatures living so close to us – not only nearby but common, for that's what the book said – that I knew nothing about. I knew nothing because I never looked, I never asked. As the Sowerby's beaked whale swims hidden beneath in the depths of the sea, so the elephant hawk-moth flies hidden beneath the depths of the night.

It was as if I had hardly noticed a bird in my life before, and had always been content with one word to cover them all. It was a bird, wasn't it? How much more detail could anybody possibly want? And then, in a single morning, being shown a swan and an eagle and a flamingo and a hummingbird and a bird of paradise. Or as if I had been content to refer to all wild mammals as 'animals': and then in the space of a single morning having an encounter with both an elephant and a tiger.

For that's more or less what happened. Down at the bottom of our garden there are elephants and there are tigers. And I never knew.

Graduation Day. Eddie was leaving school. Going to college next term. Big adventure.

The occasion was marked by a ceremony at the school. It was a lengthy business. Each leaving pupil was required to stand on stage while one of his teachers delivered a eulogy.

It was a good occasion. These teachers are good people, who take on children with all kinds of social difficulties and educational problems. Some of the children can be rather overwhelming; others have a default tendency to disappear into a silence. Eddie's first four years there had been pretty miserable. It was only towards the end that he made friends and felt supported, and that made it all bearable. Certainly, when there was any doubt about whether or not Eddie was well enough to go to school, he always and loudly cast his vote in favour of going. This was not because school was much cop: it was more about Eddie's bravery – and his strong desire not to be different from the rest. The more different he felt, the keener he was on fitting in.

His eulogy was appropriately affectionate. He was inclined to bask in the atmosphere, which was sentimental *ma non troppo*, for all the kids were good kids now. He stood there in his new suit – a dashing little number in quiet but unmistakable purple – as his teacher spoke, making a big play about the way Eddie had sorted out something to do with his school lunch. He had found a certain situation difficult, and he had taken steps and sorted it out. So far so good, but it was presented as the most marvellous thing he had done in five years of schooling.

There never was a parent yet who thought the school had done his child full justice. But all the same, here was a boy who could lunge a horse, write poems, make a damn good marmalade cake, recognise a blackbird by its song, explain the different dentition of carnivores, rodents and the blue whale, and tell you how a bat finds and catches food. As well as knowing a barn owl's power.

Those teachers are doing a job that matters and one not everyone would want to do.

But as for Eddie – well, let's just say that there's always more than you think.

Nearly packed. In a couple of days we would be going to Cornwall. Eddie was looking forward to gannets and clouded yellows, treats unavailable on a Norfolk marsh. Last-minute tasks were being done, the backlog of work finished off. It was a time of drawing a line, blotting the last page, pressing Save and pressing Send.

The start of the summer holidays feels like the end of something. It was a break from the day-to-day observance of spring on the marsh, spring turning from promise to fulfilment to autumn. It was as if we had done our job: a contract had been fulfilled; nature had done what comes naturally. Creatures of many kinds had made more, perhaps many more of their kind. That was all great: and now we move on.

But not in too much of a hurry. There was a plague of butterflies – an infestation of beauty – and it needed savouring. A brief succession of warm, still days filled the garden and the marsh with small tortoiseshells and red admirals, with large and small whites, with common blues. It was as if a stained-glass window had deconstructed itself, each separate tiny pane taking wing across the cathedral of the marsh.

In front of my hut a heron stood staring at the water, trying to look as if it knew what it was doing. It had the height and the beak and the legs, it had the fierce yellow eye: but there was a little down on the leading edges of its folded wings. It was a half-grown thing, a teenager newly fledged from the heronry trying to be a grown-up. It was the nearest a heron could ever get to looking sweet.

I suppose all wild sounds are equally wild if you think about it logically, but some sounds seem startlingly untameable. The cries of young kestrels are like that: reckless commitment to the wild air, to the life of swoop and pounce and hover. Now the young birds were working, by means of a period of loud and swaggering play, at the skills they would need to survive when they came to live by the wing. Only the best of them would survive the hard winter …

But hard to think of death when the sky was full of the wild, yelling joy of being a young kes. Life is hard, for humans and for kestrels and for everything else, but that doesn't invalidate the moments of delight that come our way. The ecological holocaust continues, we lose species, we lose abundance. We know all that, we accept all that, and most of us do at least something to make things better or to stop them getting worse – but these hard, unpleasant truths don't invalidate our moments of delight in the wild world.

A kiss from your beloved doesn't cure all the harms in the world, but for an instant it seems to.

17

THE MAD CONDUCTOR

The stable door has become a gun that fires deadly volleys of swallows.

There were rosehips along the lane. Blackberries too:

Blackberry, Blackberry, mind the prickles!

We don't like it when you tickles!

An extract from 'The Blackberrying Song', composed and performed by Eddie and me.

It had rained a fair bit while we were away on our tour of the West Country: the dykes were fuller, the reeds were taller. The marsh had been growing and changing behind my back, much as Eddie does when I'm away.

There seemed to be young birds everywhere, hopping and flitting about naively, as if they hadn't yet made up their minds about flight-distances and the trustworthiness of humans. Sometimes they came astonishingly close, as if they didn't see us as threats at all. And what's more, they were right. Eddie and I had no interest at all in catching Robin Brownbreast – a robin not yet in adult plumage – and making a meal of him. But when – if – the bird at our feet became an adult, he would be a good deal more circumspect.

Robins are more trusting than most birds: they have worked out that there's profit to be made from foraging around a working human. The robin on the spade handle is a truth as well as a cliché: a tribute to their adaptiveness. When I'm mucking out, there's generally a robin on the muck-heap, looking for scraps: when I'm sweeping the yard, there's generally a robin cleaning up the spiders.

But what if these young birds, still closer and more confiding, failed to lose this excessive trust in humankind? I can tell them with great authority that if they did so, very few people in Britain would raise a hand against them. Would the forces of natural selection favour ever-bolder robins?

You can see changes in bird behaviour where there is a local tradition of tolerance. Sometimes at an outdoor cafe, sparrows will come to the table and feed on crumbs. In Barbados I was able to lure the lovely little yellow birds called bananaquits down to my breakfast table by scattering sugar. Bill Oddie can sit in his small, gnome-thronged Hampstead garden and feed great tits and blue tits by hand: he has convinced them of his trustworthiness.

How long since small birds have been safe from humans in this country? When did we stop netting them and bird-liming them? Perhaps a century ago. But you can't explain to the birds, as Withnail did to the invading Uncle Monty: 'We mean no harm!'

Often a pair of mallards come and sit on the small pond in the garden: almost always they fly when I walk past, despite my best Withnail imitation. You can explain that, I suppose, by the fact that people still shoot ducks and do so with immense ferocity less than a mile away. But equally often, sometimes in the course of the same 50-yard walk, I pass a pheasant that scarcely bothers to step out of my path – and the destiny of this bird is noisy slaughter at human hands. So explain that.

It's true the birds on the marsh, even when they have reached maturity, are comparatively tolerant of human presence and prepared to be convinced that we mean no harm – so long as we don't get uncomfortably close. There's a small sense that on the marsh, the normal rules have changed just a little: that flight-distances can be reduced, at least to an extent.

This is a good thing for the birds: it means they don't have to burn energy on needless escapes. It's good also for humans, in that it gives us this treasured vision of Eden, the fleeting feeling that the lost paradise – every paradise is by definition lost, whether it is Eden, childhood, a holiday, a long-ago love affair, or the perfect pub – is within reach once again, and can be regained.

As birds get fewer and humans get more numerous,

perhaps evolution will favour birds who are more comfortable around humans. Perhaps we can see a future in which, with every sit on a park bench, a human will feel like Adam or like Eve: surrounded, as in the picture by Rubens and Brueghel, by the living throng. Perhaps my dreams of avian plenty are a foretaste of some future time when birds, of necessity, accommodate themselves to human life.

Of course, they already do this. The dove – the same species that appears in white form as a symbol of the Holy Spirit – has come to our cities to live alongside us . . . only to be despised and, where possible, exterminated. City pigeons, feral pigeons and all forms of racing pigeons and ornamental doves have been bred from – are – the same species as wild rock doves. The white doves in a dovecote, along with the fantails and tumblers and pouters, are all genetically rock doves.

But still, as I sat out on the marsh, relishing the start of autumn – though it was still Eddie's summer holidays – I enjoyed the notion of the robin on the table, the sparrow on the ground, goldfinches along the back of the bench, mallards in the water, an egret at the water's edge, and perhaps, on my shoulder, a singing blackcap.

I was getting foolish. As soon as a bird gets bolder around humans we invent new names for them. Pest. Vermin. We dream of a lost time when we lived at one with the wild world and non-human life: and yet when there is any chance at all of recovering even a small part of that vision, we reject it with horror and take lethal measures.

Time to get back to the house. Cook supper. Pour a drink.

Morning ride. A small green dragon flies alongside us. Not a woodpecker at all . . .

Movement, movement. It seemed that the whole world was on the move. Birds have wings and that frees them from the tyranny of place. The grass snakes of the marsh have limited options in terms of distance; some of these birds have half the earth at their command.

That restlessness, that sense of transition. Birds that had been paired-up were now forming flocks: and there was a flight of 100 goldfinches, flying over in that bouncing way that finches go in for and tinkling urgently to each other in encouragement and solidarity.

Evenings were marked by large movements of gulls flying to roost on the open water on the far side of the river. We tend to feel that black-headed gulls are beneath our notice: but those evening formation-flights would be considered a sight of breathtaking beauty, if only the birds had the decency to be rare, if only they had the decency to be more wary of humans.

I looked up as 200 of them passed together in a series of wide, shallow vee-shaped straggles. They were mostly black-headed gulls, but there were also herring gulls and lesser black-backed gulls among them: all the gulls you would expect, in short. And from the heart of them I heard – or did I imagine it? – a sort of pantomime dame's expression of mild surprise and disappointment.

Eoh!

I listened out for a repetition, but it never came. And that was mean of it because I'm pretty sure it was a Mediterranean gull. But you can't go counting every prob and poss that comes along, that's cheating. Too-light winning makes the prize light, after all. So Med-gull failed to make the list while I plumed myself on my honesty, like a golfer marking the position of his ball with fanatical correctitude.

There had been two other close calls that week. A flight of a dozen linnets – also bouncing in the finchy manner – had me checking the list. And then quite distinctly I heard the triple-note of greenshank. It's a sound I associate with the Luangwa Valley: as soon as the river level begins to fall and the beaches appear again below the banks, greenshanks come winging in, confirming their presence with that cheery piping. A bird that in this country is associated with wind-blown winter estuaries always reminds me of the belting heat of the end of the dry season in the Valley when the midday temperature is around 45 degrees.

But both these species had turned up on or over the marsh before. Lists may not be a big deal for me: but all the same, that century, achieved this year, was important. When a cricketer scores a century, the only thing that matters, in strict sporting terms, is what it does for the team, and whether or not it helps the team achieve victory or avoid defeat.

And these few acres of marsh are managed for the sake of the team, for the sake of the wildlife of the world. But all the same, the century tells us that the marsh is doing a good job. It's a deserved honour for the marsh itself.

Now, as the autumn movements gathered pace, its job was to provide winter food and shelter for those that needed such things.

Morning ritual. Nice cup of tea; great spotted woodpecker.

If you want to learn more about the wild world, don't take a field guide, take Eddie. For a start, you might not be out there at all without Eddie's insistence. The combination of his curiosity, his frequently unexpected ways of understanding what we find, and his gift for contemplation are all fine ways of approaching nature. What you see is great, but the greater thing is being out there. Not what you look at but what you're part of. And that is the greatest gift the marsh brings to us. We're not audience, we're participants.

It's like experimental theatre of the 1960s, when the relationship between cast and spectator is challenged. The fourth wall no longer exists, the proscenium arch has been torn down; this is theatre in the round, the audience promenading, joining in the songs, half-guilty witnesses of atrocity, half-courageous supporters of what's right, part-innocents caught in the crossfire, or part of the silent conspiracy of those who know they ought to speak out but daren't.

The difference is that theatre is illusion and pretence, while what happens out on the marsh is as real as anything ever could be. And you know a theatrical performance will start at 7.30pm, whereas out on the marsh the performance is endless.

If this is theatre, who is the star? You can't have a show without a star, can you? It has to be the marsh harriers: large birds of prey compel the attention more effectively than most other species on the planet. And when the bird in question has overcome extinction (or near-extinction) not once but twice in the past century, they acquire a certain added charisma. First the guns wiped the bird out altogether from England, and then the pesticides brought the bird down to that single pair. Now they cruise with the greatest nonchalance in all the watery places of East Anglia.

That distinctive silhouette has become a kind of signature: an autograph, scrawled in the sky with a mixture of flamboyance and understatement. It's a single initial, a vee, a shallow vee, and if it's not a long signature it's utterly distinctive and impossible to forge. You see that vee scrawled onto the page of the sky, forever low, for they are birds that prefer to operate at crop-dusting height. It's there, above the reeds, above the grazing marshes, the eternal, abiding image of this marsh, this part of England. There is land, forming the bottom inch or so of the picture, the reeds making their endless verticals: a thin gold line. Above them that vee, like a double brush-stroke, between the low reeds and everything else. Because the rest of the picture is sky, all sky, coloured and recoloured, reworked every day, every hour, sometimes it seems every minute, with 50,000 shades of grey and blue.

I was pretty sure that Pale had nested away to the left and Dusky on the right – Dusky with the female with green wing-tags, and Pale with the unmarked bird. And there, as Eddie and I sat on the marsh, beans consumed, jam jar empty and spoon licked, a marsh harrier went past, dramatically close – perhaps naively so. I suspect that the bird wanted to get a good look at us, to gain more knowledge about the world. What are the creatures that sit on benches and eat beans from a jar? And drink golden fluids from a bottle? I got a good look myself and saw that the bird was still a trifle fluffy: all at once it was a moment of special rejoicing. It was dark, darker than either of the adult females, its gender yet to announce itself.

Here was a young bird, then. Here was a member of this

year's crop. There had been success: there were more marsh harriers in the world than there had been a few weeks ago, and these few acres had played a small part in that process.

It was a moment of achievement: perhaps the year's greatest moment of achievement. The top predator, as we have seen, is the most vulnerable species in every ecosystem: very many things have to be right before a big bird of prey can survive, prosper and make more of its kind. When such a bird can do just that, it shows one or two good things about the environment. That it could be worse.

If the marsh had been a cricketer, it would have been entitled to remove its helmet and allow the world to admire its features, then to raise its bat to the pavilion and to the crowd, and to allow the applause to wash all around. The century of birds, with the appearance of the waxwing, was a great moment: but it was not the year's great achievement. The arrival of the singing Savi's warbler was wonderful and worth boasting about again and again, every time I had a conversation with a birder, but it wasn't the summit of the year. This was: time to look skywards in thanks, to receive the hug of the batting partner, to know the commentators are pouring praise down from on high, to know that in the press-box the journalists will be searching for the *mot juste* to encapsulate a small triumph.

Swashbuckling, they might say. Or battling. Meticulously compiled. Defiant. Effortless. Faultless. Full of murderous certainty. Chancy. Riding his luck. Inevitable. Glorious. Unforgettable. Masterful. Match-changing. Life-changing. Heroic. Reckless. Scrupulous. Filled with the perfect joy of futile heroism. Perfect for the day: knowing that other days will follow, days of ever-greater uncertainty.

So let's mix metaphors. What are metaphors for, if not for mixing together? Suddenly the marsh is no longer a cricketer, but an orchestra, and I'm the Mad Conductor. When I was in my teens I bought a conductor's baton and, alone in my room, I would conduct Beethoven's Ninth symphony from first note to last. I would bring in the flute with a wave of my little finger, silence the strings with the flat of my palm, and then, come the last movement, I would bring in the entire choir with a full swooping swish of the baton that was more Errol Flynn than Sir Malcolm Sargent, and at my express bidding, those hidden voices would sing out together just as the morning stars sang together in the great painting by Blake, and I would lead them all in a voice that attempted, in both volume and expression, to make up for its lack of tunefulness, and the choir, taking a lead from their master, would sing out from the single plastic speaker of my reel-to-reel tape recorder:

Freude! Freude!
Freude schöner götterfunken . . .

Joy! Joy!
Joy, beautiful radiance of the gods . . .

I would lead the orchestra and the choir and the four soloists to the conclusion and then I would silence them all with a final drastic razor-fighter's slash of my baton, and the world would shudder in a great seismic roar of silent applause . . .

And I, bloody fool that I was, would bow deep and take every scruple of credit for the vast and glorious din that I and our neighbours had just enjoyed. I remember going to the

Proms and standing almost in the orchestra as Colin Davis conducted the Ninth, and when it was all gloriously done, he, like me, soaked up the applause as a dry land sucks in the storm. And he had done nothing: he had made not a sound all night; the glorious din was made by people other than him, and yet he was the one who bowed.

So as the young harrier finished his inspection of the two crouching creatures below and sailed off, wobbling slightly as he sought to master the harrier's art, I received the silent applause of the world, as if I had created that bird myself, as if I had brought the two parent harriers together, Wing-tag and Dusky, or maybe Pale and Untag, as if I had personally given them food and shelter, as if I had allowed their offspring to make the extraordinary transfiguration from fluffy dinosaur to grown-up, glorious, not-quite-adult bird. And it's all true: I played as much part in the making of a marsh harrier as I did in the making of a symphony – after all, I switched on the tape recorder, did I not? And threaded the tape through the play-heads and into the take-up spool? And pressed start? Yes, I did all those things: and on the marsh, did I not press the start button, did I not raise my baton, did I not sing along with all my might? Sure, sure, I did all those things.

When the Mad Conductor had recovered from his exertions, ceased from his labour at Los's forge and was able, after this phantasmagorical outpouring of his life force, to take up day-to-day responsibilities again, I would press the rewind button. The tape that had dawdled through the play-heads while the music was going began racing with appalling speed back onto its home reel, as if to show that *ars* was every bit as *brevis* as *vita*. Uncannily soon, the leader tape would leap free and whirl around at high speed, slicking and clicking until I pressed stop and restored it to its box.

The concert was over.

But not the music. And there would be more music tomorrow.

That was, I suppose, the year's high point. Not hope but achievement. The first kiss is not the best kiss (damn it, is there really room for another metaphor in this chapter? Of course there is, bring it on), it is merely the most exciting. The best kiss is the next.

Eddie and I returned jars and bottles to the bag and made our way back to the house. Without Eddie I wouldn't have seen that marsh harrier. Or enjoyed it in quite the same way. Eddie too could raise his bat, Eddie too could bow to the audience. It was his triumph as much as anyone's.

18

HERE HARE HERE

Three fallen leaves in the water butt. Get back on the tree, you little traitors!

I bumped into Jane the shepherd as she was attending to her flock on the land next door. She told us that she was cutting down on sheep numbers, and was going to give up her tenancy. Being a generous person, she was keen to let us know; she knew we had attempted to rent the land ourselves but had lost out to her five years ago. The land is in the gift of the parish, and the trustees preferred to support a long-time local person rather than a bunch of arrivistes: no complaint from us.

But we had been here ever since, and what with the marsh and Eddie and all, it was clear that we weren't fly-by-nights or faddists. We decided to approach the parish, tell them we were interested, and see how things went. I called Helen at Norfolk Wildlife Trust, and she was naturally keen that the place be managed for wildlife. She gave us some thoughts on the best form of management: basically, to graze the sward from mid-July onwards, until the place had been eaten off.

We talked about this to Jane when next we saw her, and suggested that she might consider grazing a few sheep there herself, without charge, at a time we suggested. That looked like a good idea. Goodwill was cemented, and so – as is the alarming way of life – a fantasy seemed to be becoming a reality slightly faster than we were prepared for.

Nothing was settled, but everything looked good. There were about ten acres there, varying qualities of grazing marsh crossed with dykes and some patches of reeds. A couple of old corrugated iron pig arks had been dumped there, and would be a serious job to remove. There were also some wacky pieces of ironmongery, from long destroyed gates, looking like a mildly avant-garde sculptural installation. A few bales lay around: part of the area had been mowed and at least some of it baled. Other parts glowed bright green like a golf course. This was what's called improved grassland: treated with fertiliser and probably selective herbicide.

In other words, the place was full of – that dreaded word in conservation – potential. It was a lovely spot, from which at least some of the loveliness had gone missing. It looked to me as if it could do with a little tenderness: and the first act of kindness would be a little benign neglect. Well, I was an expert at that.

So we wrote to the parish again and said that we would love to take the place on. They said that should things go ahead, a five-year deal would be the thing, with a view to renewal after that, if all went well. This seemed good: more than good.

So as the year began to shut down, insofar as a wild year ever does shut down, so the air was filled with thrilling new possibilities for the year that would follow. We would be more or less doubling the wild land we were responsible for: and if we couldn't do anything about the long-term future of this new patch, we would be able to do something about its awkward present, if we found favour. Norfolk Wildlife Trust said they would support us. As we researched all this, it became clear that the common was already a county wildlife site – even though it wasn't a patch on our own land that lies next door. Nowhere to hide a Savi's warbler, anyway. But Helen also told us that she had done the research and discovered that most of our own bit of marsh is also a County Wildlife Site. Well, glory be.

But perhaps, if it all happened as we wished, this new patch would be jumping with orchids, once the land was grazed outside the flowering season. Who knew what lay beneath the soil, locked in the seedbank, awaiting its moment to make itself known? You don't, after all, know what happens next.

You know that spring follows winter, but you don't know what kind of spring. Perhaps it will be rotten and wet and late and horribly short on wildlife; perhaps it will come dismayingly early. Perhaps the Savi's will come and breed; perhaps the Cetti's will fail to show. Perhaps cranes will come and nest on the ground on the land next door; perhaps there will be no swallows at all. Perhaps there will be a painted lady year like 2009, when the buddleias almost broke under the weight of these lovely migrating butterflies; or perhaps it will be a wet and cold disaster.

The only truth you can be certain of is that life will do its best to keep on living: and all you can do is do your best to give it half a chance. Being where we are, we can do this in terms of land, in terms of the actual stuff that wild things live on. We can carry on looking after an island of wildlife – an expanding island of wildlife – in one of the most nature-deprived countries in Europe.

> The moon is almost full. The barn owl on his hunting perch is not.

College was good. No more school uniform. These days, three mornings a week, Eddie would set off in his leather jacket and his Elvis tee-shirt. He was given more scope, more responsibility for himself, and he took it on as a thirsty man takes on water. You change, you develop, you move on: it's a process that never stops. Even if you have Down's syndrome.

Now every Wednesday he went to Clinks Care Farm. He would come home both knackered and ravenous, scarcely able to speak. 'Are you the farmer?' I would greet him as he returned. That's one last Withnail joke, but so far as Eddie is concerned, it is a proud acknowledgement of his new identity. So what was happening today?

Feeding the pigs, he would explain. Building a bonfire. Collecting the eggs and sorting them into sizes. Looking after the shop.

The year was turning, the autumn jobs were in full and urgent spate, and Eddie was a part of it, locked into the natural rhythm of the world.

A vast bowl of pasta would be a matter of urgency, or maybe pancakes: load up those carbs. Some non-taxing television. A computer game with his generous brother. And then a vast and steaming bath because the pigs and the rest had left their mark on him, and tomorrow was a college day.

After that, hair soft and smelling of shampoo, in the contentment of weariness, I would read to him from a mildly testing tome of natural history, and we would discuss such important matters as why toothed whales differ from baleen whales, and then it would be goodnight.

A last goodnight from Cindy. She came downstairs – Eddie apparently asleep before she reached the landing – to report that he had said something significant.

'I love my life.'

Morning ride. Almost under my horse's feet, two baby stoats playing chicken.

Roger Wilson was a great conservationist. He was never one to brag about it much, and his name is not a famous one; he just did immense amounts of fantastic work. He worked with Dian Fossey and the gorillas in Rwanda, he worked in many other places in Africa, and latterly he worked with World Land Trust and was a mentor and an inspiration to the new generation of conservationists. He was a visionary and a wholly practical man; he was an idealist and a cynic. And above all he loved the wild.

A casual caller would have assumed his job at World Land Trust was sentry. I'm a council member as well as a long-time supporter of this fine organisation; I often dropped in to their headquarters in Halesworth, and most times I found Roger on guard outside: smoking a cigarette with the air of a man doing his duty. Whenever we had a pint in the Angel, he would take out his tobacco pouch and roll cigarettes: for the pleasantly therapeutic nature of the task, and as if he was planning to smoke them all one after another as soon as he got outside. Alas, the fags got their own back, and he died from lung cancer: death followed diagnosis with horrible rapidity, and there was devastation and dismay throughout World Land Trust.

His life was formally celebrated by a gathering at the Royal Geographical Society in London, which lies hard by the Royal Albert Hall. I was asked to speak. Honoured, I prepared my brief talk. It was when I started practising it that I realised I had written it as a joke with a punchline: therefore I needed the nerves of a stand-up comedian to deliver it properly. And that's not what I'm best at. When it comes to public performance, I make up for lack of flair by conscientious hammering and copious rehearsal.

But I remembered that when Cindy worked as an actor, she had been in a farce called *They Came from Mars and Landed Outside the Farndale Avenue Church Hall in Time for the Townswomen's Guild Coffee Morning*. It played for three weeks at the Pleasance at the Edinburgh Festival and toured for a year afterwards. Cindy worked on her comic technique with the play's author, David McGillivray, who was called, in one review, 'the master of timing'.

One line of Cindy's consistently got a laugh, but not as much as David believed it should have done. Cindy was encouraged to take a beat – a pause – between lines and the joke did better. She then tried two beats – and each night she got a real shout of laughter. But give it more – she never did – and it would get no more than a titter, David said. So I bore that in mind when I went up to the podium at the RGS. Here's the first chunk of that address:

> Roger was a wild man. Wildness flowed in his veins and filled his heart. In a suit he looked like a man togged out in fancy dress; in a room he looked like something in a cage. Lord knows I can tell a few tales of the wilderness myself but, without ever trying to be competitive, Roger could beat me hands-down every single time. One tale alone shall have to do for the lot: and it tells of a wildness so deep that it is obvious that Roger had gone a million miles beyond puerile ideas like trying to conquer the wilderness, impose his will on it or use it as a proving ground for his manhood.
>
> The tale involves a little bit of African lore, so I'd better give you the footnotes before we start. It's about a long-drop. This, as all old Africa hands know, is a camp latrine. The name gives you a pretty good idea of the science behind it, and you will easily understand that the longer the drop, the better it is for all concerned.
>
> I have encountered long-drops with decadent luxuries like a sawn-off oil-drum fitted with a real lavatory seat, but the hard-core tend to shrink in horror from effete

nonsense. Such people have always preferred the simple cross-pole. You will easily understand that it's a good idea to choose a stout pole, and then to anchor it reasonably securely. There are plenty of tales of those who neglected such precautions.

The second footnote involves the spitting cobra. This is a genus of venomous snakes with a dozen or more species in Africa and Asia.

These catch their prey by the approved serpentine method of injecting it with venom, but they also use venom for a rather drastic defence mechanism, one that they exploit when cornered. They turn themselves into a highly dangerous water-pistol and shoot their venom at whatever has disturbed them. They can fire a remarkable quantity of the stuff for a good couple of metres.

You're unharmed, if somewhat startled, if the venom merely lands on your unbroken skin, but the cobras tend to go for the eyes and they're pretty good shots. A direct hit can cause temporary and permanent blindness.

Roger and I were swapping Africa stories. He told me of a camp he had lived in for some time. 'That's where we had a spitting cobra in the long-drop.'

Well, what would you do? Would you kill it? Would you take the more risky strategy of removing it – and be forever after fearful that it might come back to a place it clearly had a fancy for? Or would you request others to get on with the job while you had urgent business elsewhere? Well, Roger – what did you do?

One beat. Two beats.

'Wore goggles.'

I got my shout, or rather, Roger got his. So we raised glasses – rather than goggles – to Roger, and vowed that if we wouldn't see his like again, quite in that way, we would do our best to make sure that we saw something reasonably similar, at least in terms of commitment and ability.

Nicola Davis – author of *Bat Loves the Night* and *Big Blue Whale*, already quoted in these pages – is a trustee of World Land Trust as well as a friend of the family. She called the occasion 'a rededication'.

I felt the same thing. Sometimes a couple choose to renew their wedding vows; the people who take part in the first Mass of Easter renew their baptismal vows. It's not that you had forgotten them, or set them aside, or wilfully broken them: it was more an acceptance – or perhaps a realisation – that you believe in the things you had made your vows for more than ever. And more than ever, you wanted to keep doing what you believed was right.

So yes, as the year on the marsh – a year since we had taken on those extra acres – came to a conclusion, it was a realisation or an acceptance that managing a piece of land for wildlife was a good and right thing to do. We must carry on doing all we can to keep the place singing.

It was the same with the stuff I was writing. The loss of biodiversity and bioabundance is a greater threat to the planet than climate change. But as I write my stuff, I must make it clear that fragility and importance are only part of my subject – for the wild stuff out there is also wonderful beyond description. As we live with less and less non-human life in our lives, so our lives become poorer. We are increasingly living in a monoculture. That not only impoverishes us: it makes us less human.

So, rededicated, I resolved, among other things, to write this book as well as I could, so that I could tell the world about gorillas, spitting cobras, marsh harriers, Savi's warblers, bats, Sowerby's beaked whales, bumblebees, hawk-moths and bonking beetles. Without such things, we are less than ourselves.

Julian Roughton from the Suffolk Wildlife Trust paid us a visit. He and I had been looking at the incredible Carlton Marsh project: one that involves more than 1,000 acres of marsh, with scrapes and pools and dykes and bunds and miraculous volumes of Broadlands sky: and all of it filled with birds and the cries of birds. It was half an hour's drive from our place: not much more as the harrier flies. It was a little more connectivity for nature in our nature-deprived land, and I was thrilled by the place for itself – and also thrilled by the fact that it was Julian's personal achievement. I had written to him when the purchase was completed: 'I know you will tell me that it's all about the team, and that you've got a great team and that nothing could have been done without them, and of course you're right. But there comes a time in a football match when a striker scores the winning goal. At that moment he is entitled to acknowledge the applause of the crowd: applause that's for him alone. So perform your knee-slide, remove your shirt and wave it above your head: because you have done something great.'

So yes, Julian is one of the really good guys – and a man worth listening to, a man of easy manners and open mind. So Cindy confessed to him the worries we often felt about managing the marsh, and how so many times we felt we had fallen short.

'Look, the reason this place is special is because no one has done anything with it for years,' Julian said. 'It's fabulous. And it's OK to let it get on with doing whatever it's doing. Don't worry about the list of jobs. They're just a guideline, a perfect-case scenario. It's a wild place: if you just let it get on with the job of being wild, you're doing a good job yourself.'

> Listen while I tell you about the hare in the garden. All ears?

The hare in the garden was no longer a stranger. He had become a local character, nothing less. It was clearly the same hare seen many times, rather than a series of different hares. We had watched him grow from a leggy leveret to a slightly less leggy adult. He seemed content with a solitary life: and I wondered if or rather when that would change. Come spring, would his urge to be with others of his kind prompt him to move back to the uplands? Or would wanderlust drive others down towards this fringe habitat? For this wasn't classic hare country. These days they exploit big open areas of farmland, preferring dry country and light soils. But down here on the marsh, they had a place without huge vehicles, without much disturbance, without dogs, and where the only workers were us. No one would be shooting at them. The hare was certainly not considering these matters: but he had found a place where he could make a decent living and was in no itching hurry to leave.

And as for us ... well, we were sharing our living space with a hare.

Hares have their witchy side, like all creatures of the

night, for they are mostly nocturnal. All the same, I see half a dozen of them most days when I ride out: generally they cower under our hooves, breaking cover at the last minute to streak away across the field. I have always loved that moment when they raise themselves up on these long levers, when they change from possible rabbits into definite, undeniable, for-all-time hares. It's that transition from bunny to antelope that I take most delight in.

But as the hare grew accustomed to the garden and to the drier half of the marsh – the now dog-free half we bought from Barry all but 12 months back – I was beginning to see a different kind of animal. Not the back end of one running away, nor the humpty shape of a hare pretending to be invisible: but a hare loping at ease, at peace with the world – and by implication, at peace with me. With all of us.

He had worked out that headlong flight from the humans he shared the place with was uneconomical: burning up energy to no purpose. So he recalibrated his ideas of what constitutes a safe distance; so long as he kept to it, he was content to feed and rest without excitement, and with us in plain view maybe 20 yards off.

A hare's vision is a colossal thing. Those enormous golden eyes on the side of its head give it 360-degree spherical vision: he can see danger from above and all around without needing to move his head. So he would crouch nibbling with his back half-turned towards me, knowing exactly where I was and what direction I was moving in. If I approached a yard or so nearer – sometimes inevitable if I was walking the wriggling path around the edge of the marsh – he would rise up on those long limbs and move in what we horsemen called a collected canter – a million miles from the headlong gallop both the hare and I know pretty well – and shift another five yards away. Then he would stop and wait and maybe feed, and if the path took me away or allowed me to keep the same distance, he would remain as he was, aware but not exactly wary. Here was a hare who did not feel the need to go haring off.

In a small way, he had changed his relationship with humanity. And because we had been looking after the marsh and spending time there, we had changed our relationship with the hare. New limits had been established, new boundaries had been drawn, a new treaty had been recognised. What was obviously true of the hare was more subtly true of many other creatures that lived there. Inside our boundaries, the rules had changed.

It wasn't quite the Eden of the painting, when Eve could have stretched out a naked arm to stroke a peacock or a tiger, but the hare was closer than he was six months ago; the other creatures that lived here were often a yard or two closer than they had been when we first took the place on. We came without threat, we walked slowly, we sat quietly. Like Withnail, we meant no harm: and that message had – at least to an extent – got through and been accepted. We felt content and at ease in this small patch of lovely land: it was good, and more than good, that some of the non-human creatures that we shared it with felt ever so slightly the same. So one more line from *Withnail* to celebrate:

Here hare here.

Easy to make too much of this. But when you have shared a place with the hare that stayed, you feel ... well, you feel that something has been exchanged.

Morning ride with butterflies. The comma has yet to bring summer to a full stop.

'Chiquitito!' Ruby implored. 'Chiquitito, come and talk to us!'

And the grey whale, apparently responding to Ruby's voice, came still closer, approaching our small boat in the San Ignacio Lagoon in Baja California in Mexico. Soon, astonishingly, he was hard up against the side of the boat, and recklessly we leant out and offered pats and strokes to the rubbery skin, while the whale, twice as long as the boat, seemed to get an odd pleasure from this strange moment of contact.

Eddie liked the story. He had heard several times about the whale that came up to be stroked and the lady who called him *chiquitito* or little darlin'. I told him again about the whales, and how they actively seek out these encounters in a place where they were once hunted to bloody death, and how it felt to lay hands on such immense creatures of such immense wildness.

One of the reasons he likes the story is because he likes Abba, especially the songs used in the musical *Mamma Mia*.

Chiquitita, you and I know

How the heartaches come and they go and the scars they're leaving ...

So we sang as much of the song as we could remember. We sang for the whales, for Ruby, for tales and events and creatures of immense wildness. We were out in the marsh, both of us dressed in waterproof tops and waterproof trousers. The rain had eased off, though it was not the sort of weather you would normally choose for a picnic. I had tried to get out of it and, as usual, failed. Eddie was mildly surprised that

I should have been anything less than eager to undertake a picnic in the rain.

The attentive readers will be able to predict what happened next: cheese and tomato sandwiches, baked beans in a jar, fruit and yoghurt in another jar, apple juice, beer. OK? That's everything. No, you can carry the bag.

So, game as ever, he swung the bag onto his shoulder and we set off. One of those strange facts of life: it's always raining harder when you're indoors than when you step outside. It wasn't as bad as I feared, but then it hardly ever is. Dusk wasn't far off. Well of course, it wasn't: this was, I remembered, the September equinox. But a good hour of light remained, maybe a little longer, despite the cloud. A year ago – a year ago almost to the day – we had learned about our acquisition of the new section of marsh, had listened for bats, had talked about echolocation.

We sat on our favourite bench and I unpacked the treats. It's surprising how pleasant it can be, drinking beer in the rain: getting wet on the inside and the outside at the same time. As always, Eddie ate with great concentration. Above us black-headed gulls were flying towards the open water of the flood. Heading north.

It was then that I was aware of a counter-movement taking place a little lower in the sky. Swallows? No, the shape wasn't quite right. Then I heard the call, like someone blowing a raspberry, affectionately rather than derisively. House martins then. Half a dozen, no more, a dozen – more than two dozen. Eddie stopped spooning beans to watch as they approached us from the north, circling and circling, spiralling and figure-of-eighting, jinking in the air when one of them found a morsel of aerial plankton to snack on. Southing, southing.

'Where they going, Eddie?'

'Don't know.'

'Yes, you do. Af—'

'Africa!'

'Yes! Brilliant boy! And that long journey, do you know what it's called? Mi—'

'Migration!'

Each time one of those martins jinked, it took on a morsel of protein: a morsel of marsh. A long, long journey: and we had made it the tiniest bit more possible.

These long-distance migrants' once-brilliant strategy has made them vulnerable in the shrinking world. They need more than just a few acres to live their lives: they need acres and protein spread out over many thousands of miles. Swallows, swifts and martins: they'd struggled this year. And perhaps will struggle the next and the one after that. Vulnerable.

Perhaps you think I'm being needlessly glum. If you love wildlife, you know about gloom, know about it with immense precision, but you don't dwell on it. Neither despair nor mad optimism is much help. It's about doing what you can to make things a little better.

Chiffchaff, willow warbler, Cetti's warbler, whitethroat, blackcap, garden warbler, sedge warbler: they had all bred in and around the marsh in the course of the year. And two pairs of marsh harriers had nested nearby, using the marsh for hunting, and there were certainly young ones fledged. Vulnerable species in a vulnerable world: but there seemed to be some local success here. And the otters, the mysterious otters – they were still about and leaving their secret messages here and there: the occasional spraint on the path, the slide that leads from the bank into the dyke as well maintained as ever. Cindy was making a fine piece of work with swimming otters. Joseph was making music; teaching music too, these days.

Eddie, the most vulnerable one of us, had moved from the beans jar to the pudding jar. The rain was gentle now: a subtle freckling of the waters of the dyke. Beneath our layers we were still warm enough. No need to hurry. It was getting dark; the house martins, moving on, would soon be roosting somewhere south of us. Perhaps already. Good luck on that long journey.

What of tomorrow? What of next year? And after that, what then for the vulnerable?

What indeed.

It was time to go in, but then again, it wasn't. Eddie had slipped into an evening reverie: life around us teeming away, subtly, quietly, eternally. The darkness grew: it seemed to rise up from the ground inch by inch. A small yelp, like someone treading on the paw of a Yorkshire terrier.

'Little owl,' said Eddie.

Beneath our feet the world turned and kept turning.

Epilogue

The Marsh

Dad saw a deer
in the distance
the river
looked colourful
and pretty
I saw a butterfly
on the beautiful flower
we saw a barn owl fly
around the marsh
the herons
were chattering
in the heronry
I can hear trees
whooshing
in the wind
birds
are all singing
the marsh
full
of life

the end

The Bird List

2012–18*

Mute swan
Pink-footed goose
Greylag goose
Canada goose
Egyptian goose
Shelduck
Wigeon
Gadwall
Teal
Mallard
Shoveler
Pochard
Tufted duck
Goldeneye
Red-legged partridge
Grey partridge
Pheasant
Cormorant
Little egret
Great white egret
Grey heron
White stork
Little grebe
Red kite
Marsh harrier
Hen harrier
Sparrowhawk
Buzzard
Kestrel
Hobby
Peregrine
Moorhen
Coot
Crane
Oystercatcher
Lapwing
Snipe
Woodcock
Curlew
Common sandpiper

* On 2 January 2019 a flight of seven spoonbills crossed the marsh.

Green sandpiper
Greenshank
Redshank
Black-headed gull
Great black-backed gull
Common gull
Lesser black-backed gull
Herring gull
Common tern
Rock dove/feral pigeon
Stock dove
Wood pigeon
Collared dove
Turtle dove
Cuckoo
Barn owl
Little owl
Tawny owl
Swift
Kingfisher
Greed woodpecker
Great spotted woodpecker
Magpie
Jay
Jackdaw
Rook
Carrion crow
Goldcrest
Blue tit
Great tit
Coal tit
Marsh tit
Skylark
Sand martin
Swallow
House martin
Cetti's warbler
Long-tailed tit
Chiffchaff
Willow warbler
Blackcap
Garden warbler
Lesser whitethroat
Whitethroat
Savi's warbler
Sedge warbler
Reed warbler
Waxwing
Treecreeper
Wren
Starling
Blackbird
Fieldfare
Song thrush
Redwing
Mistle thrush
Robin
Dunnock
House sparrow
Grey wagtail
Pied wagtail
Meadow pipit

Chaffinch
Brambling
Greenfinch
Goldfinch
Linnet
Bullfinch
Reed bunting

Plant List

From survey performed by Norfolk Wildlife Trust,
17 July 2017

Wild angelica
Sweet vernal grass
Fool's watercress
Lesser burdock
False oat-grass
Daisy
Downy birch
Grey willow (sallow)
White bryony
Purple small-reed
Hedge bindweed
Lesser pond sedge
Greater tussock sedge
Knapweed
Common mouse-ear
Creeping thistle
Marsh thistle
Hazel
Hawthorn
Cock's foot
Common couch
Great willowherb
Hemp acrimony
Meadowsweet
Cleavers
Ground-ivy
Reed sweet-grass
Hogweed
Yorkshire fog
Hop
Frogbit
Iris
Sharp-flowered rush
Toad rush
Compact rush
Soft rush
Hard rush
Blunt-flowered rush
Meadow vetchling
Common duckweed
Least duckweed
Honeysuckle
Greater bird's-foot trefoil
Yellow loosestrife

Purple loosestrife
Black medick
Watermint
Water chickweed
Watercress
Reed canary-grass
Common reed
Ribwort plantain
Rough meadow-grass
Silverweed
Tormentil
Common fleabane
Creeping buttercup
Yellow rattle
Bramble
Common sorrel
Clustered dock
White willow
Grey willow
Elder
Water figwort
Ragged-robin
Bittersweet
Prickly sow-thistle
Marsh woundwort
Hedge woundwort
Lesser stitchwort
Dandelion
Common meadow-rue
Red clover
White clover
Common nettle
Brooklime
Pink water-speedwell
Tufted vetch